Horst Piller

MICROSCOPE PHOTOMETRY

With 79 Figures

Springer-Verlag
Berlin Heidelberg New York 1977

DR. HORST PILLER
c/o Carl Zeiss
Postfach 35/36
D-7082 Oberkochen

ISBN-13: 978-3-642-66572-1   e-ISBN-13: 978-3-642-66570-7
DOI:10.1007/ 978-3-642-66570-7

Library of Congress Cataloging in Publication Data. Piller, Horst. Microscope photometry.
1. Microphotometry. I. Title. QH257.P54.   578'.4.   76-58893.

Softcover reprint of the hardcover 1st edition 1977

2132/3130-543210

Preface

Over the last dozen years there has been a great development in the study of matter on the micro-scale by means of the light microscope, which has coincided with the rapid growth of automation and the use of computers. The manufacturers have responded to this challenge, and there is now available a large choice of modules for use with the microscope, both on the biological and non-biological side.

This book is on microscope photometry. Photometry on the macro-scale can be applied to features of not less than about one millimetre in size. Below this lies the realm of microscope photometry which will be discussed both in theory and in practice. As a general rule no discussion of technical design of equipment is included, as such information is available in the handbooks of the particular manufacturers of equipment. It attempts to cover all the uses of the optical microscope photometer with its auxiliary equipment, and the first chapter describes the topics covered. In using the word 'light' we apply its meaning to extend from UV to near IR.

No attempt has been made to list all the works consulted. For each subject treated in this book a list is given of books and papers considered to be of key importance; in these will be found more detailed bibliographies.

It is from Carl Zeiss (Oberkochen, West Germany) that this book has sprung; the author would like to express his personal gratitude to Dr. Michel, Dr. Weber and Mr. Muchel. In the work of publication the author is particularly indebted to Dr. K.F. Springer and the staff of Springer-Verlag. Dr. N.F.M. Henry of the Department of Mineralogy and Petrology, University of Cambridge has freely given advice and help with the English text.

Oberkochen, April 1977   HORST PILLER

Contents

CHAPTER 1

General Introduction

The essentials of the microscope have been known since the time of Abbe, but over the last dozen years or so development of new and improved apparatus has been rapid. The use of computers for the design of lenses, the production of a range of very sensitive photosensors, and the growth of automation have been the main factors in this development. We are now able to measure the intensity of light from areas only one micrometre in diameter, to produce many data and to process these, all in a short time. Thus the time has come for a description of microscopy today and for an assessment of its present capabilities and potentialities. The present book is a contribution to this.

Specialised instruments have been developed for particular purposes with names such as microphotometer, microspectrograph, microdensitometer and others, but we shall not be concerned with these in the present book. Each has its own design, and reference should be made by the user to handbooks of the manufacturers.

Our concern will be entirely with the ordinary microscope to which can be added various mechanical, optical and electrical modules for special purposes and in particular for the measurement of light in transmission intensity and in reflection, including interference techniques and the measurement of fluorescence. Such an instrument is best called a microscope photometer.

There is a fundamental difference between macro- and microscopic photometry; in the former, light from the object is measured directly, while in the latter, it is the light from an image of the object that is measured. This underlines the requirement for an image of the best possible quality, that means a good microscope properly used. Accordingly, we shall study first the basic microscope and the adjustment of all of its parts and the functioning of all the modules (mechanical, optical, and electrical) that may be used along with it. The range of electromagnetic energy that will concern us is that with wavelengths from 250 to 1100 nm, so that in referring to 'light' we mean to include ultraviolet (UV) and near infra-red (IR), as well as the visible range which we take for ordinary purposes to lie between 400 and 700 nm.

Our field of study will go only as far as the obtaining of good data in its final form. Further subjects related to the data obtained, such as the area representation of absorption (pattern recognition) and its various aspects in the field of biology or the identification of solid substances as well as the correlation of data with submicroscopic structures in this material will not be our concern. Automation of measuring procedures and of data processing is described only briefly in Section 9.6. The additional instrumentation, special sample preparation and operation involved in automation are so voluminous that a separate book would be necessary for this subject. However, routine use of automation is only possible as yet in a few cases and the efficiency of automation is improving so rapidly that it seems advisable to publish such a book at a later date.

In a book such as the present one a decision has to be made about how far to go in explanations. Here the aim is to explain all the essentials of the phenomena and of the correct adjustment of the apparatus. Full use is made of diagrams with detailed legends, and these are employed for explanations, instead of text, in all cases where a two-dimensional lay-out is better fitted for the purpose. For further details and explanations reference is made to other books.

The main divisions of the book and the contents of each chapter are set out below.

Ray Paths and Energy Transfer. Our first concern (Chap. 2) is with understanding the geometry of the microscope. This involves an appreciation of the reciprocal relation between field and aperture area and of the process of image formation. The various kinds of illumination are studied.

In Chapter 3 we study the transfer of light energy in the optical system as a whole. This is of fundamental importance because light is a form of energy that reacts with matter and that is recorded by an electronic photosensor or by a photographic film, in which case it has to be described either radiometrically or photometrically. It is also a stimulus to the eye, which is a psychophysiological apparatus responding to the visible range only.

Detailed Description of the Equipment Parts. In Chapter 4 we treat the stabilised light source and the ways in which the light beam from it can be modulated.

The subject of monochromatising devices is discussed in Chapter 5, including interference filters, polarising and other filters; these modify the beam from the lamp on its way up to the photosensor (including the eye).

In Chapter 6 we come to the basic microscope itself, and all its parts are treated: condensers, reflectors for work in reflected light, stages, objectives, body-tubes, and oculars.

Finally we deal with the photometric system (Chap. 7), and study the types of photosensor and their electrical characteristics, including those of the further modules required to concert the output in the way most suited for our purpose.

Experimental Adjustments and Procedures. Having finished the detailed study of the various parts of the equipment, we take up the study of the experimental adjustments and procedures. In Chapter 8 we adjust the modules of the basic microscope, including those required for work with birefringent substances. Chapter 9 is occupied with procedures for photometry with the microscope, both with biological and with non-biological specimens. In Chapter 10 the subject is that of special techniques in microscope photometry that have not been covered in Chapter 9. Lastly in this group of chapters comes the matter of the special adjustments for photometry with the microscope (Chap. 11).

Errors. Two chapters are devoted to errors and their reduction within tolerable limits. Chapter 12 deals with systematic errors arising from modules or their adjustment. The other kind of error is that called statistical arising from the inevitable distribution of measuring values about a mean (Chap. 13).

Applications. The last chapter (Chap. 14) deals with the applications of microscope photometry in various fields of biological and non-biological science.

Appendixes. The appendixes cover certain items of theory and of numerical calculation, along with some tabulated data. One of the most important appendixes is that dealing with symbols and defining equations for all the terms used in the book. Only terms approved by some accepted authority have been used, and every effort has been made to be consistent in terms and symbols.

The quantitative derivation of colour parameters from spectral-distribution curves is not included in this book although this subject has, in the last years, become of growing importance for the identification and study of microscopic objects. Reference may be made by the reader who is interested in this topic to the relevant literature on colorimetry, e.g. Billmeyer and Saltzmann (1966), Hardy (1936), or Henry and Phillips (1977).

The apparatus used by the author in experiments described in Chapter 11, Sections 1 to 5, in the Appendixes 4 and 6, and in the legends of Figures 6.8, 9.5a, 10.5, 11.1, 11.3, 11.5, 11.6, 12.4, A2, A4 and A5 have been made or supplied by Carl Zeiss, Oberkochen, West Germany, unless another manufacturer is indicated. This equipment was most conveniently available for the author; it can be taken as being of standard design so that the test results may be achieved as well with equipment from other manufacturers.

CHAPTER 2

Ray Paths in the Microscope

2.1 Introduction

The descriptions of ray paths in the microscope given in most books are insufficient to provide what is required for a proper appreciation of the roles of the various parts in a modern microscope. This chapter attempts to deal with ray paths in enough detail to illustrate the concepts of geometrical optics and the interrelation of lenses and diaphragms. The essential optical system described here exists in every microscope, but some are equipped with extra parts, optical and mechanical, for the measurement of optical and geometrical properties in various kinds of stage objects.

Since the subject of this chapter is largely geometric, the details are mostly given in the figures and their legends. The text proper provides a framework for these and also gives the more general statements. The concepts, terms and symbols are set out so as to obviate the need for their repetition in the legends. Conjugate planes in the light train are those in which the same image is formed, back or forward; hence, mention of a given plane involves also all those conjugate to it, without these having to be specified at each mention.

For brevity, the terms objective, condenser, collector, illuminator, and microscope are used for the lens in question as well as for the whole body in which the lens is incorporated. The term lens is used for both a single lens and a lens system with specified function. The name diaphragm is applied to what actually is the opening of a diaphragm.

The subject of the transport of light energy is covered in Chapter 3, but this chapter provides the geometric basis for that subject. The subject of the diffraction theory of the microscope is omitted. On this subject the relevant literature may be consulted, e.g. Michel (1964).

The stage object may be illuminated with different arrangements of optics, but in this book we shall use only the arrangement called 'Köhler illumination.' The features of this are as follows:

1. The luminous surface of the lamp is imaged into the plane of the condenser-aperture diaphragm, so that adjustment controls the aperture angle which can be reduced below its maximum.

2. The luminous-field diaphragm is placed in a plane where a back image of the stage object is formed, so that its adjustment limits the area illuminated on the stage.

3. As a consequence of these two basic features, it follows that the stage object outside the illuminated area neither scatters light nor does it suffer possibly harmful effects through heating (which can happen with biological specimens).

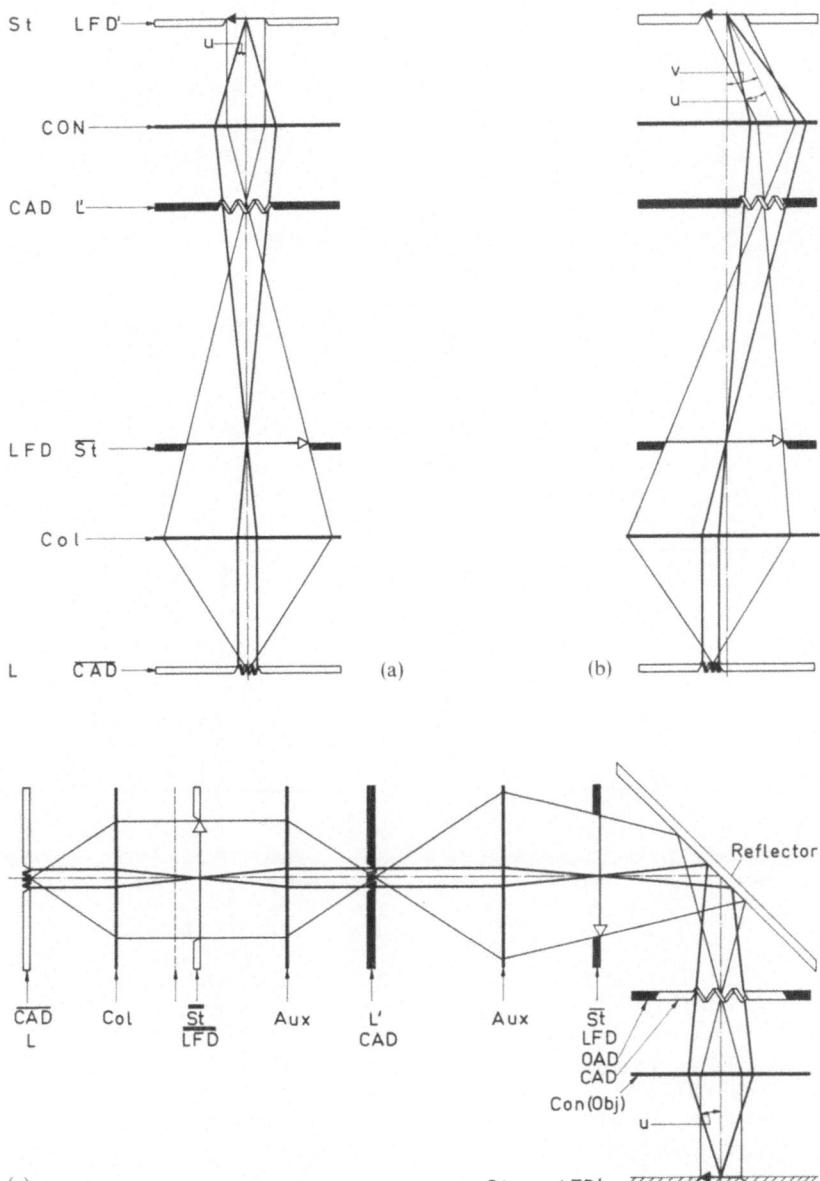

Fig. 2.1 a–c. Legend see pages 8 and 9

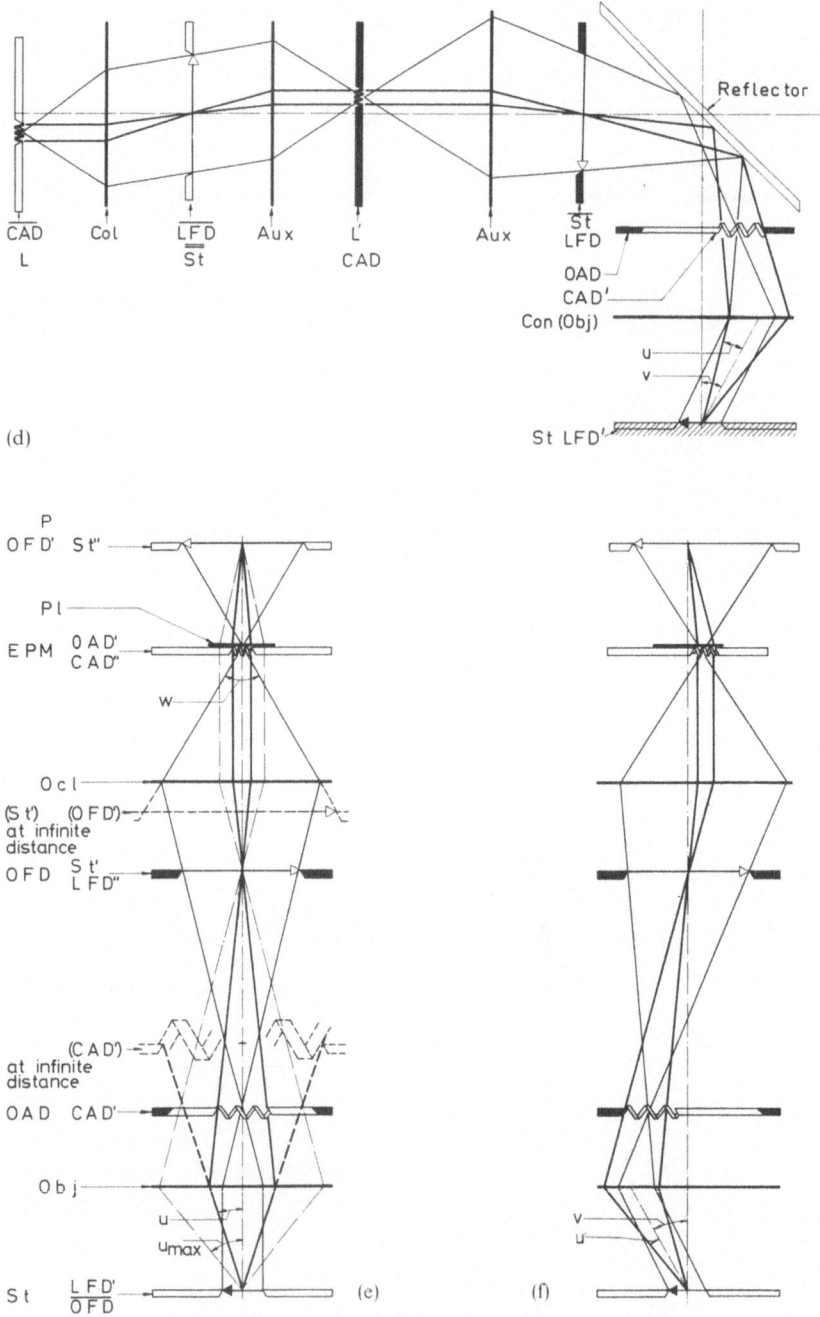

(d)

(e)

(f)

Fig. 2.1 d–f. Legend see pages 8 and 9

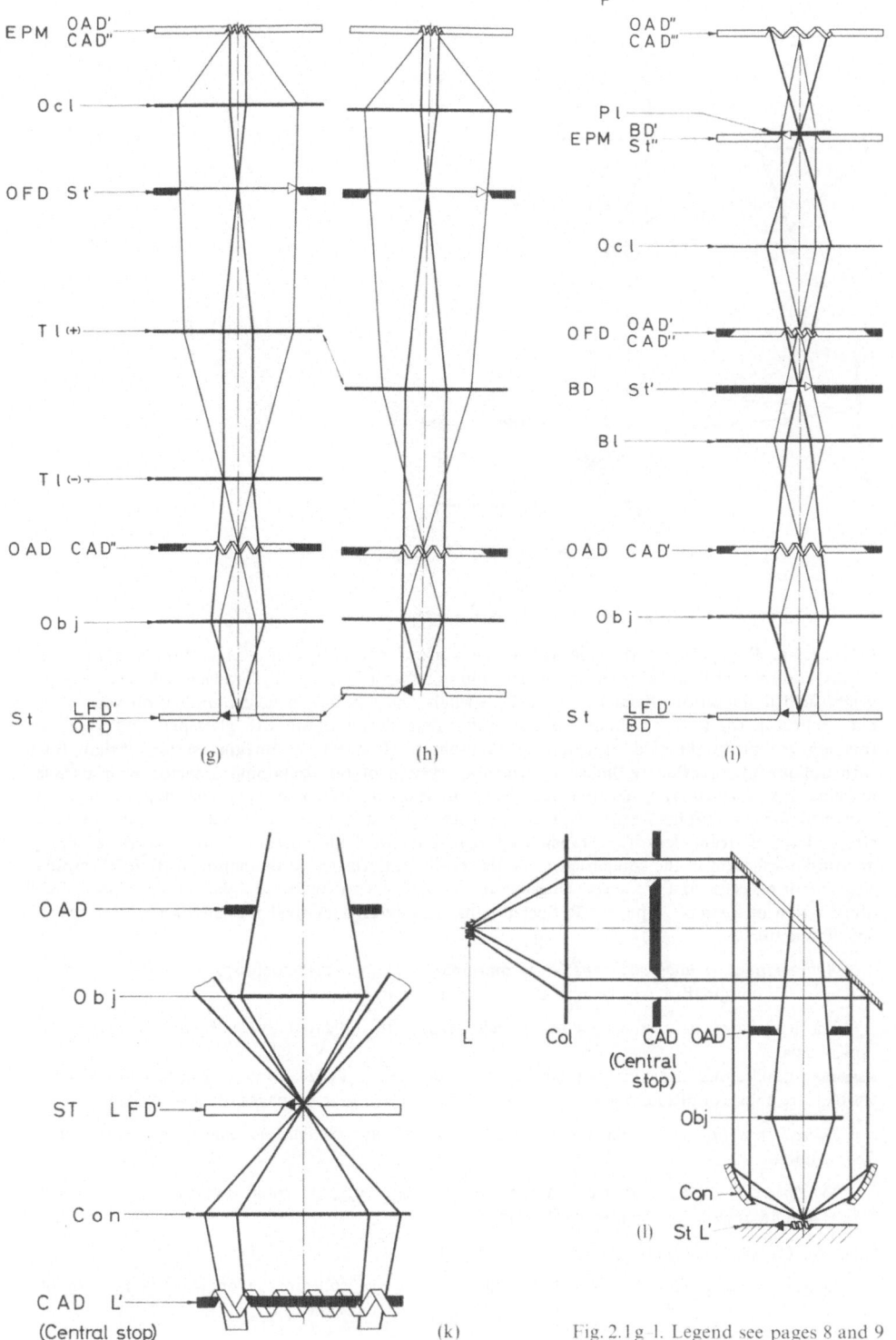

(g)   (h)   (i)

(k)   (l)

Fig. 2.1 g–l. Legend see pages 8 and 9

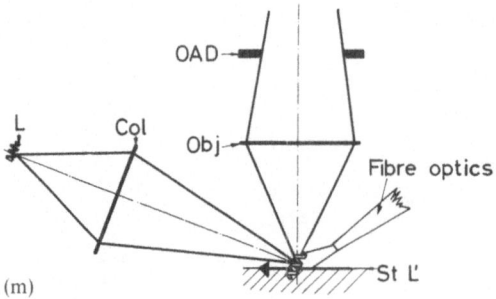

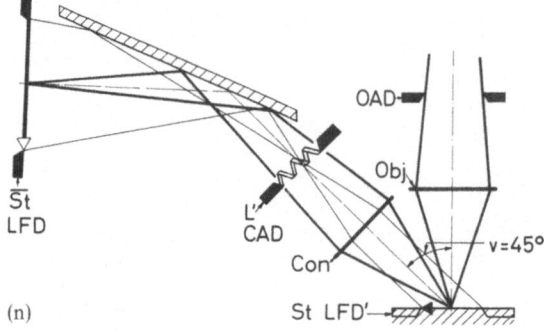

Fig. 2.1a–n. Ray paths in the microscope for various conditions of illumination and imaging.
(a) Transmitted-light, normal bright-field illumination (Köhler). (b) Transmitted-light, oblique bright-field illumination (Köhler). (c) Reflected-light, normal bright-field illumination with glass-plate reflector. (d) Reflected-light, oblique bright-field illumination with glass-plate reflector. (e) Imaging system, bright field with normal illumination (Köhler). (f) Imaging system, bright field with oblique illumination (Köhler). (g) Imaging system of (e) containing a sector with parallel marginal rays formed by a negative tube lens. (h) Imaging system of (e) containing an objective corrected for an infinite image distance. (i) Imaging system, normal bright field, showing the effect of the Bertrand lens. (k) Transmitted-light, dark-field illumination with a central stopping aperture-diaphragm in the condenser. (l) Reflected-light, dark-field illumination with an ellipsoidal ring-mirror reflector and a concave ring-mirror condenser. (m) Reflected-light, unilateral dark-field illumination of a simple type. (n) Reflected-light, unilateral dark-field illumination with a special 45° illuminator.

Abbreviations, symbols and signs: —— marginal rays; —— principal rays; –·—·—·— optical axis; .... marginal rays forming maximum aperture angles.

The lenses or lens systems including field lenses are indicated by *thick bars* normal to the optical axis.
▰▰▰▰▰▰▰ diaphragms; ▭▭▭ real images of diaphragms; ⋀⋀⋀ luminous surface of the lamp; ⋀⋀⋀ real images of luminous surface; ◂– stage object; ◂–real images of stage object.

Dotted straight lines: a virtual image or a virtual ray. *Oblique hatching below or behind a line:* a mirror.

Symbols: u effective aperture angle, object side; u_{max} potential aperture angle, object side, w field angle; v obliquity of illumination.

Note: For u, v, and w see also Figure 2.2.

Lenses: Aux auxiliary lens; Bl Bertrand lens; Col collector; Con condenser; Obj objective; Ocl ocular; Pl photosensor lens including the lens of the eye; Tl($\pm$) tube lens, + positive, − negative.

Also, the field areas are uniformly illuminated, and light scattering from lens mounts and tube walls is much reduced.

Note: Figures 2.1a and e show that in the transmitted-light case the geometry of rays in the illuminating system is the same essentially as in the imaging ray system, if we go away from the stage object in both cases. Thus, the stage object can be thought of as an inversion point about which the microscope system is symmetrical. It follows that we could take the imaging systems of two microscopes and use one of these as the illuminating system of the other by placing the luminous surface of the lamp in the same plane where the system has its exit pupil, when it acts as imaging system. For illumination, the objective becomes the condenser and the ocular becomes the collector. In practice, however, other specially-designed lenses are used in the illuminating system; these need not be so well corrected as those of the true imaging system; but they do enable the operator to vary, within large limits, the diameters of fields and apertures by convenient adjustment of the diaphragms, while the imaging-system diaphragms must be operated and lenses exchanged for this purpose.

◀ Fig. 2.1 a – n (continued)

Optical elements: EPM exit pupil of the microscope; L luminous surface of the lamp; L' image of the surface; P photosensitive layer (retina of the eye, photocathode, photographic film, screen); St stage object.

Diaphragms: BD Bertrand diaphragm; CAD condenser-aperture diaphragm; OAD objective-aperture diaphragm; LFD luminous-field diaphragm; OFD ocular (field) diaphragm; PFD photometric-field diaphragm (Photometer diaphragm).

Main marginal ray set for bright-field illumination	Main principal ray set for bright-field illumination (transmitted light)
Luminous-field diaphragm (LFD)	Luminous surface of the lamp (L)
Stage object (St)	Condenser-aperture diaphragm (CAD)
Primary image (St')	Image of condenser-aperture diaphragm (CAD') in the objective-aperture area
Retina or screen (P)	Exit pupil of the microscope in the pupil of the eye (CAD'' $\cong$ EPM)

For reflected light there are two images of the condenser-aperture diaphragm in the objective-aperture area so that CAD' and CAD'' are in the same level and the exit pupil of the microscope is CAD''' $\cong$ EPM.

Notes:

1. Abbreviations primed (e.g. St', L', LFD' etc.) are forward images, while those with a bar over (e.g. $\overline{\text{St}}$, $\overline{\text{CAD}}$ etc.) are back images. The number of primes or bars indicates the order of the image within the sequence of images.

2. With specularly reflected light the order number of the images of CAD or L in the imaging optical train is one more than indicated in the figures.

3. The broken vertical straight line with the arrow in Figure (c) denotes the place where an auxiliary lens for adapting a monochromator (Fig. 5.4) may be located in the ray path

2.2 Illumination Systems for Bright Field

2.2.1 Normal Incidence (Fig. 2.1 a)

If we consider a pencil of rays diverging from a point on the luminous surface of the lamp, we can see that the collector[1] will 'collect' these and bring them again to intersect at a certain distance — thus forming an image of the luminous surface. In this plane is placed the condenser-aperture diaphragm. The rays that intersect in the centre of this diaphragm are called principal rays. These rays are made parallel by the condenser so that they traverse the stage object as a cylindric bundle, and so delimit the illuminated part of the stage object. On the other hand, if we consider a bundle of parallel rays coming from the luminous surface of the lamp along the axis of the optical system, we see that these will be brought by the collector to a focus at its focal plane on the side towards the condenser. In this plane is placed the luminous-field diaphragm because the optical arrangement is going to form the image of the stage object here if the latter would be illuminated from the side of the objective. Instead of being parallel, the rays coming from the luminous surface of the lamp can converge and intersect just in front of the collector towards the condenser, and the luminous-field diaphragm is placed just in front of the collector. Taking this pencil of rays, we see that these are marginal to the aperture area of the condenser which brings them again to a focus, this time at the stage object. Half the width of the aperture area dictates half the angle subtended between two opposite marginal rays in the stage object, and this is the aperture angle.

This is the essential illuminating system for transmitted light, although it may be elongated by auxiliary lenses to produce a section in which the marginal rays are made parallel before the condenser is reached; this allows the insertion of plane optics, such as filters and polars without displacing the image of the luminous-field diaphragm.

For reflected light, however, auxiliary lenses are necessary for the following reason. Here the objective has to do double duty, acting as the condenser in the illuminating system as well as the objective in the imaging system. Thus, restriction of a diaphragm in its focal plane on the side where the light comes from would constrict also the objective-aperture area; as this would not be tolerable, a system including auxiliary lenses has to be adopted. The illuminating system for reflected light (Fig. 2.1 c) thus contains an additional plane of the aperture-area type; an additional image of the luminous surface of the lamp is formed and in the plane of this image is placed the condenser-aperture diaphragm. The reflector, which causes the horizontal beam to be deflected downwards on to the stage object, can be placed at any level in the body tube above the objective and below the ocular, provided that it is of the plane-glass type. A prism reflector, on the other hand, must be placed as near as possible

---

[1] This module is also called the 'lamp condenser,' but the term used here is preferable in order to avoid confusion with the stage condenser.

Fig. 2.2a–d. Cones of light made visible in opalescent glass. Light cones accepted by the objective (aperture cones).

(a) Vertical illumination with full aperture angle (u_{max}) of 23°. (b) Vertical illumination with aperture angle restricted to 11.5°. (c) Illumination at an obliquity of 11.5° and an aperture angle of 11.5°.

Note: In all three of the figures the stage object lies at the shared apex of the pair of light cones.

(d) Light cone above a high eye point ocular (viewing-field cone). The total field-of-view angle (*w*) is 30°, and the apex of the cone is at the exit pupil of the microscope

to the aperture area at the back of the objective so as to avoid unilateral shadowing of the primary image in the imaging system.

The part of the whole reflected-light microscope lying between the lamp and the objective acting as condenser is called the Reflected-light Illuminator (German: Auflicht-Illuminator) and it comprises the following: condenser-aperture diaphragm, luminous-field diaphragm, reflector and auxiliary lens or lenses.

2.2.2 Oblique Incidence (Fig. 2.1 b and d)

One-sided incidence is used for visual observation as it produces relief, which can make a detail in the stage object visible that might not be noted in normal illumination; in photometry, errors arise from obliquity of incidence (Sect. 11.5). The condenser aperture diaphragm is displaced laterally so that, although remaining normal to the optic axis of the microscope, it is no longer centred to this axis. This causes the principal rays to pass obliquely through the stage object and so to intersect in off-centred points in the aperture areas. Although the apex of the cone of marginal rays remains in the centre of the stage object, the axis of this cone makes an angle with the optic axis of the microscope called the obliquity of the illumination.

Note: Since the lenses are unaltered in position, the off-centering of the condenser-aperture diaphragm causes its back images to delimit a different area of the luminous surface of the lamp; where the surface is small, this may entail moving the lamp laterally. The obliquity can be measured along with the aperture angle, so that both can be properly controlled (App. 14).

2.3 Imaging Systems for Bright Field

2.3.1 Ray Paths Without the Bertrand Lens (Orthoscopic Imaging) (Fig. 2.1 e)

The marginal rays from a point in the stage object are brought by the objective to intersect again in the plane of the primary image which lies in the front focal plane of the ocular. All marginal rays diverging from a point in the primary image are made parallel by the ocular. When these rays fall upon the eye accommodated to infinity the lens of the eye then forms the secondary image of the stage object on the retina where they intersect again. The parallel rays coming from the ocular are seen by the eye to form a virtual image of the infinitely-distant primary image. But for calculation of the angular fields, the distance from the stage object is taken as 250 mm, this being the least distance of distinct vision (or conventional viewing distance). The lens of the eye may be replaced by a positive glass lens (photosensor lens) that forms the real image on a photocathode, a screen or a photographic emulsion (in which case it may be called camera lens or projecting lens).

The second possibility for forming the second real image is to displace the ocular towards the observer along the optic axis of the microscope; this causes the marginal rays to intersect and form the image. Alternatively, the distance between the stage object and the objective can be increased so as to form the primary image below its proper position; this, however, disturbs the quality of the image and renders the set of objectives no longer parfocal.

A third way of forming a second real image is to use a special negative projecting lens in place of the ordinary ocular; this has the advantage that, with a small number of lenses, an image of excellent quality can be obtained. This kind of arrangement is often used for projecting the image on a screen, or on to a photocathode or photographic film.

The principal rays that will form a pencil leave the stage object parallel to each other and are made by the objective to intersect in its back focal plane, whence they diverge to the rim of the ocular diaphragm. The ocular then makes these rays intersect in its focal plane (observer side), which is, in effect, the focal plane of the microscope as a whole; it is called the exit pupil (Ramsden circle) of the microscope and it lies a few millimetres above the top of the ocular at a distance convenient for the position of the pupil of the eye. The area of the pupil of the microscope has to be smaller than that of the pupil of the eye, otherwise the observer will not see the whole field. A photocathode may be placed here, provided that only a small area has to be irradiated.

The ray paths so far described do not allow for the insertion of plane optics by which we mean filters, polars, beam splitters, etc. as these would alter the level of the primary image. To allow their use without this bad effect we must make use of tube lenses that will produce a certain distance of parallel path in the marginal rays. This can be accomplished either by means of a negative tube lens (Fig. 2.1 g) or by using an objective corrected for infinity (Fig. 2.1 h); in both cases a positive tube lens must also be used in order to form the image at the required plane. If these tube lenses do alter the size of the primary image, we need to know the magnitude of the 'tube factor,' and this is explained in Chapter 6, Section 6.

Of course, tube lenses also change the paths of the principal rays, so that the exit pupil of the microscope is moved closer to the ocular below it; this requires the use of an ocular having a higher eye-point, otherwise the surveyable field would be reduced.

2.3.2 Ray Paths with the Bertrand Lens (Conoscopic Imaging) (Fig. 2.1 i)

The (Amici-) Bertrand lens is inserted in the tube in order to enable the observer to see the aperture areas without having to remove the ocular. This lens causes the marginal rays to intersect, thus forming the primary image below the plane of the ocular diaphragm; at the same time it makes the principal rays intersect in the plane of the ocular diaphragm. Thus, there is a mutual change in the levels of the two respective sets of images. In the exit pupil of the microscope there is now formed the second image of the stage object. In the plane of the new primary image of the stage object the Bertrand diaphragm can be inserted so as to restrict the effective area in the stage object. It is useful to be able to focus the Bertrand lens so as to obtain a sharp image of any feature in the aperture area; this is applied, for example, in the test for the accurate levelling of a polished surface on the stage to be studied with reflected light (Chap. 11, Sect. 2) or in the determination of angles between the microscope axis and certain rays in the stage object (App. 14). On a microscope not possessing a built-in Bertrand lens, a separate device for doing the same thing can be inserted in place of the ocular (auxiliary microscope or pinhole diaphragm).

2.4 Dark-Field Systems

In a dark-field system no ray from the illuminating system is allowed to enter the imaging system, either by transmission through the stage object or by specular reflection from it. Only light scattered from the stage object reaches the eye, so that the features observed consist of small inhomogeneities in the specimen. Dark-field illumination can be achieved in several ways, but it is not possible to control the light so closely as can be done with bright-field illumination. In consequence the stage object is usually not evenly illuminated, even where the arrangement is symmetrical. This unevenness can be reduced by placing a ground-glass screen between the collector and the central stop diaphragm. Dark-field illumination can be achieved in the following ways:

1. In transmitted light a central stop in a ring diaphragm in the condenser-aperture area produces a circular cone-shaped beam with its apex at the stage object (Fig. 2.1 k). The cone formed on the far side is directed outside the cone accepted by the objective, so that the latter receives only light scattered by the stage object. This requires that the image of the filament formed in the condenser-aperture area must be as wide as the outer diameter of the ring diaphragm. Such an image can be formed from a small lamp area by means of a special condenser having reflecting concave and convex ring surfaces (e.g. a cordioid condenser).

2. In reflected light this symmetrical illumination requires an elliptical ring mirror in the illuminator, supplemented by a condenser consisting of a concave ring mirror (sometimes having a ring lens) at the periphery of the objective in order to direct the light on to the stage object at the required angle (Fig. 2.1 l).

3. Dark-field illumination can be achieved by one or more lamps (Fig. 2.1 m), each with its own collector directing a beam on to the stage object; the illumination can be unilateral, or symmetrical. This, however, demands a certain free working distance between the stage object and the objective mount, but both transmitted and reflected illumination can be achieved in this way and with any required angle of incidence.

4. In recent years fibre optics (Smith, 1975) have become important in the achieving of dark-field illumination; the exit of a bundle of light-conducting fibres being placed at the periphery of the objective and the axis of the fibres is directed towards the stage object. The arrangement of several fibres around the objective is called ring-illumination device.

5. For the measurement of diffuse reflectance under standard conditions of illumination a special type of unilateral 45°-reflected-light illuminator is used (Fig. 2.1 n). In this the optical system is similar to that for transmitted-light bright field illumination, with the condenser close to the objective, which itself is an ordinary one. Both the luminous-field diaphragm and the condenser-aperture diaphragm act in the usual way.

2.5 Diaphragms

A diaphragm is a mechanical restriction of an opening — either fixed or variable; if variable it may consist either of a set of fixed holes or an iris that can be adjusted to any required size. There are three main types: field diaphragms and aperture diaphragms have already been noted, but there is a third type of a fixed nature used for the suppression of stray light in the lens mounts and the tubes situated between lenses in the imaging and illuminating optical system.

The condenser-aperture diaphragm controls the effective maximum angle of incidence of the light cone of illumination; when the diaphragm is wide open, the effective aperture angle will be the same as the nominal one, which is the maximum possible for the lens. On the imaging side the objective-aperture diaphragm is usually a fixed one, but in certain special high-power objectives it can be an iris; this controls the maximum angle of inclination to the optic axis of the microscope of a marginal ray that can pass through the objective.

The luminous-field diaphragm controls the area that is illuminated on the stage object. The ocular (or photometer) diaphragm controls the area observed (or measured) on the image of the stage object.

2.6 Instrumental Parameters

In this section we consider the basic parameters of the microscope.

Limit of Resolution. The limit of resolution is the smallest distance between two points that can be seen to be separate in the stage object. In Abbe's theory of image formation this distance is given by the following expression (Michel, 1964),

$$d = \frac{\lambda}{(NA_{Obj} + NA_{Con})} \tag{2.1}$$

d smallest resolvable distance (in μm)
λ wavelength (in μm)
NA numerical aperture (numeral)
Obj   objective (potential)
Con   condenser (effective)

Depth of Focus (Focal Depth). There are two depths of focus, one on the side of the image, the other on the side of the stage object. That on the side of the image is the length along the optic axis of the microscope by which, for a constant focusing position of the objective, the level of the image can be varied without disturbing the sharpness of the image of a point in the centre of the field. That on the side of the stage object is the length along the optic axis of the microscope by which, for a constant level of the image, the focusing position of the objective can be varied without disturbing the sharpness of the image of a point in the centre of the field. Unless otherwise indicated the latter depth of focus applies in this book. It is given by the following equation; its first term is a geometrical optical one, while the second is a wave-optical term (Michel, 1964),

$$DF = \frac{1\,000\ \mu\mathrm{m}}{7 \cdot NA_{\mathrm{Obj}} \cdot m} + \frac{\lambda}{2(NA_{\mathrm{Obj}})^2}$$ (2.2)

DF depth of focus (in μm)
λ wavelength (in μm)
m total magnification factor (numeral)
NA_{Obj} potential numerical aperture of the objective (numeral)

Notes:

1. The depth of focus on the image side is approximately the m^2-fold of that on the stage-object side.

2. The total magnification factor may be either angular (virtual image) or linear (projected real image).

3. The expressions for the limit of resolution and depth of focus are only approximations since certain other factors are also involved, including psychological and physiological ones. For this reason somewhat different expressions will also be found in the literature (Françon, 1961).

Range of Useful Magnification. The range of useful magnification is the range of total linear magnification of the image (viewed through the microscope or projected on a screen) within which details are clearly and comfortably seen, and it (m_u) is given by (Michel, 1964),

$$500 \cdot NA_{\mathrm{Obj}} \leqq m_u \leqq 1000 \cdot NA_{\mathrm{Obj}}.$$ (2.3)

When the magnification is less than the lower limit it is insufficient to show all the detail in the image. When the magnification is grater than the upper limit it is 'empty' since there is no more detail to be revealed, and further the focal depth and the free working distance will be unnecessarily small, which can lead to a lessening of image quality through wrong cover-glass thickness and defocusing. However, experience shows that an image on which photometry is being done can be magnified up to three times the upper limit, provided that only the measured portion is homogeneous.

Magnification Factor. The total magnification factor is the product of the magnification factors of all the lenses: objective, tube lens, ocular field lens, ocular eye lens and any auxiliary lenses (in projection, also the projecting lens). In modern microscopes the magnification factor of each lens is marked on its mount, but in any case the total magnification can be measured by means of a stage micrometer and a micrometer scale placed in the image plane.

Diameter of Ocular Field. For observation, the ocular diaphragm should be as wide as possible because it is always best for the eye to observe a large area of the stage object, and ordinarily a maximum diameter of 20 mm is provided. With plan objectives and a highly-corrected ocular of the wide-field type, the diameter can be up to 30 mm. For oculars with a magnification number greater than 10, the maximum diameter of the diaphragm is dictated by the maximum field angle that the observer can view conveniently, which is about 50°, and the relation is

$$\text{diameter (in mm)} = \frac{2 \cdot \tan\left(\frac{w}{2}\right) \cdot (250 \text{ mm})}{MN_{\text{Ocl}}} \tag{2.4}$$

$w/2$ half the field angle
MN_{Ocl} the magnification number of the ocular (numeral)
For $w = 50°$ this equals $235/MN_{\text{Ocl}}$ (in mm)

Note: When the ocular has a field lens in front of the diaphragm on the objective side, the diameter is evaluated by the field-of-view number (Chap. 6, Sect. 5), which takes the place of the left-hand side in the above equation.

2.7 Definition of Geometrical Optical Terms in Alphabetical Order

Aperture
   A. angle   Half of the angle between two opposite marginal rays in a point on a field (or window) or half the plane angle of a light cone having its apex in a field (Fig. 2.2a, b).

   A. area   Dictates the aperture angle. This term is specified by the lens to which it applies.

   A. diaphragm   Mechanical limitation of the aperture area.

   Effective A.   Refers to the aperture angle that is effective during practical work; for bright-field illumination the size of the angle is the same as that supplied by the condenser; it cannot exceed the full angle of the objective. For dark-field illumination or imaging of self-luminous stage objects the effective aperture is always the same as the full one of the objective.

   Full A. (or potential)   Refers to the maximum aperture angle which the objective is able to supply.

   Numerical A.   Product of refractive index of the medium (N) in which the aperture angle is measured and the sine of the aperture angle; ($N \cdot \sin u = NA$). This term is mostly referred to the lens to which it applies, as, for example, when we speak of the numerical aperture of a particular objective or condenser.

Back image	An image that would be formed by an optical system if the direction of illumination and light-propagation is inverted.
Back side of an optical element	The side towards the image.
Central-stopping diaphragm	Stops a central portion of the aperture area of the condenser so that the angle of incidence on the stage object is larger than the full aperture angle of the objective.
Conjugate planes	Planes in which an image of the same object appears, including the plane of the object itself.
Diaphragm	Mechanical limitation of an aperture area; a field or opening in a body through which light is traversing. It may be fixed or variable in size and shape; mostly circular.
Field	Area in the level of the stage object or its images limited by a field diaphragm or the image of a field diaphragm dictating the field angle. The term is specified by position or function (e.g. ocular field, photometric field).
F. angle	Angle between two opposite principal rays in the aperture area; special case: field-of-view angle which is the field angle in the exit pupil of the microscope during observation (Fig. 2.2d).
F. diaphragm	Mechanical limitation of the field
Luminous F.	Dictates the area that is illuminated on the stage object.
Field-of-view number	When the ocular diaphragm is situated in front of the ocular (on the side of the objective) this number is the diameter (in mm) of the diaphragm, when the diaphragm is situated between the field- and the eye lens of the ocular it is the diameter of the virtual image of the diaphragm.
Forward	Refers to the direction along the optical axis on the side of the photosensor.
Illuminating optics	
for transmitted light	Comprise the collector and condenser.
for reflected (or incident) light	Comprise the collector, illuminator, and condenser, which, for bright-field illumination, is the same as the objective.

Illuminator	Comprises the condenser-aperture diaphragm, luminous-field diaphragm, reflector, auxiliary lens.
Image-side	Refers to rays traversing the primary image.
Imaging optics	Comprise the objective, the ocular (or projecting lens) and the tube lens.
Lens	The name is applied to a single lens as well as to a lens system
Amici-Bertrand L.	Lens to be combined with the ocular in order to enable the observer to see a magnified image of the aperture area of the objective.
Auxiliary L. (or relay L.)	Connects the ray paths in several optical systems or corrects aberrations.
Field L.	Lens positioned in or close to a field that is to be imaged. The field lens makes the principal rays intersect into a more convenient plane for working either of an aperture diaphragm or else a photosensor, including the eye pupil. It also makes the image of the aperture area of a certain lens to be positioned and have a size which coincides with the aperture area of the subsequent lens. A field lens has no more than a weak influence on the size, position, and quality of the image of the stage object.
Projecting L.	Lens forming a second or higher-order real image on a solid surface (e.g. photocathode, screen photofilm, etc.).
Tube L.	Lens situated in the microscope tube and suppressing disturbances in the path of marginal rays due to optical or mechanical elongation of the optical train, or lens acting together with an objective corrected for infinity in order to form the primary image.
Magnification	This term is applied if 'angular m.' or 'magnification scale' need not be specified.
Angular M.	Ratio of the viewing angle under which a distance in a virtual image normal to the viewing direction is observed through a lens to the viewing angle under which the corresponding distance in the object is observed with the naked eye at a distance of 250 mm.
M. scale	Ratio of a distance in a real image to the corresponding distance in the object.

Maximum angle of incidence	Angle between the optical axis of the microscope and the marginal ray of the illuminating pencil that is most strongly inclined to the axis, or sum of obliquity and effective aperture angle.
Object side	Referring to rays traversing the stage object.
Obliquity of illumination	Angle between the microscope axis and the axis of the illuminating pencil with the apex in the stage object (Fig. 2.2c).
Optical axis (Microscope axis)	Straight line connecting the centres of lenses.
Plane optics	Optical bodies having flat surfaces on the sides where the light enters and leaves the body (e.g. filters, prisms, etc.).
Pupil	General term for an opening that dictates the aperture angle; aperture diaphragm or real or virtual image of an aperture diaphragm.
Entrance P.	Dictates the aperture angle on the object side.
Exit P.	Dictates the aperture angle on the image side. The entrance is always the image of the exit or vice versa.
Exit P. of the microscope	Dictates the diameter of the light beam in the level where the pupil of the eye has to be placed (Ramsden circle); this is several millimeters above the ocular.
Primary image	First real image formed by the objective with or without a tube lens at a given level in the microscope.
Ray	Geometrical abstraction; a straight line which determines a direction in a light train.
Axial R.	Connects the centres of lenses or fields or pupils. If necessary the element involved must be specified.
Marginal R.	Joins the centre of a field or window to the margin of the next aperture area or pupil.
Principal R.	Joins the center of an aperture area or pupil to the edge of the next field or window.
Reflector	Part of the illuminator; it deflects the light coming from the lamp towards the stage object.
Bright-field R.	Glass plate or prism making the light traverse the objective acting as the condenser.

Dark-field R.   Elliptical ring mirror making the light pass by the periphery of the objective and impinge on the stage object.

Stage object   The object placed on the microscope stage in order to be imaged. If necessary, specified by 'specimen' which is the object to be investigated, and 'reference material' which is the object used as comparison material.

Window   General term for an opening that dictates the size and shape of a field; field diaphragm or real or virtual image of a field diaphragm.

Entrance W.   Delimits the field in the stage object.

Exit W.   Delimits the field in the image. The exit is always the image of the entrance or vice-versa.

CHAPTER 3

Transfer of Light Energy in the Optical System

3.1 Introduction

This chapter describes the factors that influence the light beam in the optical system. Knowledge of these factors is required in order to enable the operator to control the beam in such a way as to obtain the largest possible ratio of signal to noise for particular objects and photometric procedures.

For qualitative work such as observation and photomicrography, sufficient energy is usually available so that the optimum brightness and exposure time can be achieved for the particular photo-emulsion, taking into account the relation between image and background; the required information is available in the appropriate literature. On the other hand, in photometric work, in order to control the geometry of the light beam and be able to reproduce the operation conditions exactly, we need to know quantitatively the power and spectral composition of the light and the geometrical properties of the beam under the conditions of measurement.

We start with the definition of the quantity that causes the photocathode to give a photocurrent; then, we describe the factors affecting this quantity, and we finish this chapter with a short description of the relation between this quantity and the corresponding brightness to the human eye.

The photoelectric sensor transforms light energy into electric current (or any quantity proportional to an electric current). But our quantity, which is called 'radiant flux,' is really a power, that is, energy per unit time. It is symbolised by (Φ) and is expressed in watts (W).

Five factors determine the radiant flux impinging on the surface of the photocathode:

1. The initial power of the lamp, called spectral concentration (or spectral density) of radiance (B_λ) (in $W \cdot cm^{-2} \cdot sr^{-1} \cdot nm^{-1}$). (For steradian sr, see Fig. 3.1 b).

2. The spectral bandwidth ($\Delta\lambda$) (in nm, Chap. 5, Sect. 2).

3. The volume of the optical system through which the beam has passed named optical flux (OF) (in $cm^2 \cdot sr$).

4. The total transmittance of the optical system (τ_{Os}) (fraction). (For distinction between total and internal transmittance see Appendix 2.4).

5. The interaction of the light and the particular stage object (IF) (fraction or specified quantity).

Note: Factors 1 and 2 are sometimes condensed to a magnitude called radiance (B) (in $W \cdot cm^{-2} \cdot sr^{-1}$), for which the spectral bandwidth has to be specified.

The fundamental law of photometry with the microscope can be set out as follows,

$$\Phi = B_\lambda \cdot \Delta\lambda \cdot OF \cdot \tau_{Os} \cdot IF. \tag{3.1}$$

(For more information about quantities, units and symbols see Appendixes 2.4 and 2.5.)

Note: There are measuring devices which operate with light energy (quantity of light) and this is converted into an electric charge. For this case the photometric law must formally be modified so that both sides of Eq. (3.1) are multiplied by a given time and the radiant flux is converted to radiant energy.

Equation (3.1) is true for the radiant flux within a wavelength interval which is sufficiently small for the factors on the right side to be considered to be constant over that interval.

Where any of the factors on the right side of Eq. (3.1) varies noticeably with wavelength, over the given spectral interval then the radiant flux is determined by an integral expression within the limiting wavelengths (λ_1 and λ_2);

$$\Phi = \int_{\lambda_1}^{\lambda_2} B_\lambda(\lambda) \cdot OF \cdot \tau_{Os}(\lambda) \cdot IF(\lambda) \cdot d\lambda. \tag{3.2}$$

The meaning of λ as subscript or in brackets is explained by the word 'spectral concentration' or 'spectral distribution' respectively in Appendix 2.5.

In practice the integral is determined approximately. One way of doing this is to subdivide the interval into small sections of equal width; the central wavelengths of these sections are taken to represent the sections and the corresponding radiant fluxes are summed according to the equation

$$\Phi = \sum_{\lambda=\lambda_1}^{\lambda_2} B_\lambda(\lambda) \cdot OF \cdot \tau_{Os}(\lambda) \cdot IF(\lambda) \cdot \Delta\lambda. \tag{3.3}$$

The measurements that concern us in this book are based on the comparison of two radiant fluxes, one resulting from the interaction of the light with the measured specimen and the other from the interaction of the light with the reference material which, by some people, is called the standard. Thus the measuring procedure is expressed in the equation

$$\frac{C \cdot IF_{sp}}{C \cdot IF_{ref}} = \frac{\Phi_{sp}}{\Phi_{ref}} = \frac{IF_{sp}}{IF_{ref}} \tag{3.4}$$

$C = B_\lambda \cdot \Delta\lambda \cdot OF \cdot \tau_{Os}$ instrumental constant (in W)
IF Interaction factor [in terms of measuring value according to Eq. (9.1)]
Φ radiant flux (in W)
sp specimen
ref reference material

Of course, the constant (C) must be the same for both measurements in order to give Eq. (3.4) and the operator must ensure that this demand is fulfilled.

From Eq. (3.1) we draw the conclusion that, for a given interaction factor between light and stage object, the radiant flux can be systematically controlled by means of control of the quantities involved in the constant (C): the spectral concentration of radiance of the lamp, the spectral bandwidth, the optical flux of the microscope and the transmittance of the optical system. We may now proceed to study these quantities.

3.2 Spectral Concentration of Radiance of the Lamp

This quantity is characteristic of the particular lamp. An element of the area of the luminous surface emits light within a solid angle and a certain spectral range according to the formula

$$B_\lambda \equiv \frac{\Phi}{F \cdot \omega \cdot \Delta\lambda} \tag{3.5}$$

Φ radiant flux emitted by the luminous surface (in W)
F area of luminous surface (in cm^2)
ω solid angle of emitted light cone (in sr)
$\Delta\lambda$ spectral bandwidth selected by the monochromatising device (in nm)

Hence this light is expressed in $W \cdot cm^{-2} \cdot sr^{-1} \cdot nm^{-1}$. This will comprise radiant flux per unit area, unit solid angle and unit spectral bandwidth. This quantity will depend on where the central wavelength of the spectral band is taken in the whole spectrum of light emitted by the lamp. For a given adjustment of lenses and diaphragms the radiant flux supplied by the lamp is proportional to the size of the luminous surface selected by the condenser-aperture diaphragm, the solid angle selected by the collector (light-gathering power of collector) and the spectral bandwidth selected by the monochromatising device.

3.3 Spectral Bandwidth

In the preceding section we stated that the radiant flux is directly proportional to the bandwidth. But for a broad spectral band we have to take into consideration the variation of spectral concentration of radiance with wavelength; thus we express the radiance of the lamp as

$$B = \int_{\lambda_1}^{\lambda_2} B_\lambda(\lambda) \cdot d\lambda \tag{3.6}$$

B radiance of the lamp (in $W \cdot cm^{-2} \cdot sr^{-1}$)
$B_\lambda(\lambda)$ spectral concentration of radiance for the wavelength λ (in $W \cdot cm^{-2} \cdot sr^{-1} \cdot nm^{-1}$)
λ_1, λ_2 limiting wavelengths
$d\lambda$ spectral interval (in nm)

In practice the spectral interval is usually chosen to be small enough to permit use of the approximation

$$B = \sum_{\lambda = \lambda_1}^{\lambda_2} B_\lambda(\lambda) \cdot \Delta\lambda. \tag{3.7}$$

For operation with a monochromatising device the pass-band of which depends on the width of the light beam, we have also to take into account the interrelation between bandwidth and optical flux. This happens with a prism- or grating monochromator. In the case where the effective light beam has constant height in the direction perpendicular to that of the dispersion, the optical flux is proportional to the width of the beam along the dispersion direction, and thus to the bandwidth. In the case where the width of the effective light beam is dictated by a square or circular diaphragm, the optical flux is proportional to the square of the beam width, and thus to the square of the bandwidth. (For the theoretical explanation see Chap. 5, Sect. 4.3.)

Notes:

1. It is important to realise that, except where a specific loss of light occurs, the spectral concentration of radiance is the same in the lamp surface and in all fields and aperture areas. This is due to the constancy of optical flux throughout the whole light train.

2. When light is propagated in a medium of refractive index $N \neq 1$, the original solid angle in air is replaced by the numerical solid aperture, which is the product of the solid angle and the square of the refractive index of the medium (Fig. 3.1); the term B/N^2 is called the reduced radiance.

3. Since the unit of $(cm^2 \cdot sr)$ is that of optical flux (Chap. 4, Sect. 1) the radiance can be considered as light power per unit of optical flux.

3.4 Optical Flux

3.4.1 The Nature of Optical Flux

Other terms used for optical flux are: throughput (Meyer-Arendt, 1972), optical extent, light conductance. It expresses the ability of an optical system to transfer light energy, but without considering loss of light by absorption, reflection, scattering and diffraction (Hansen, 1950; Klaunig, 1953). It is, thus, a purely geometric quantity applied to the volume through which the light is transferred; the volume concerned is often called the 'light tube' (Zimmer, 1970) because it generally has the shape of a tube with circular cross-section. We shall confine our study to a light tube of this nature, because this is the case for the microscope, but a brief mention will be made of other shapes. We proceed to study the terms involved in the concept of optical flux (Fig. 3.1).

These terms are primarily the areas of diaphragms (or images of diaphragms) forming the ends of the tube which is completely filled with light, the distance between them and the refractive index of the medium through which the light is transferred.

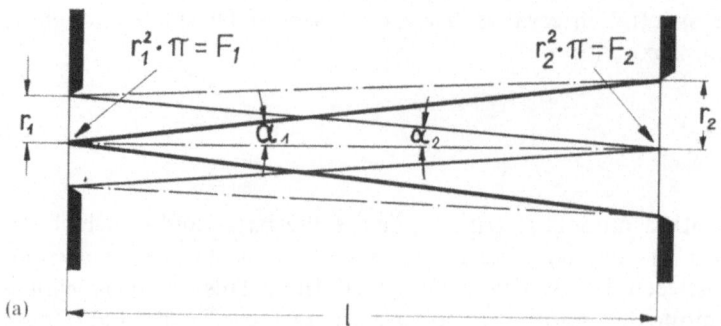

(a)

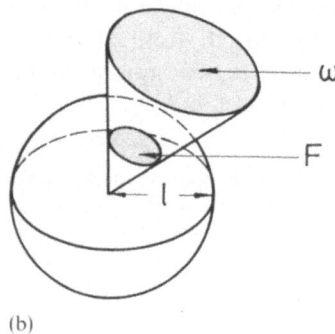

(b)

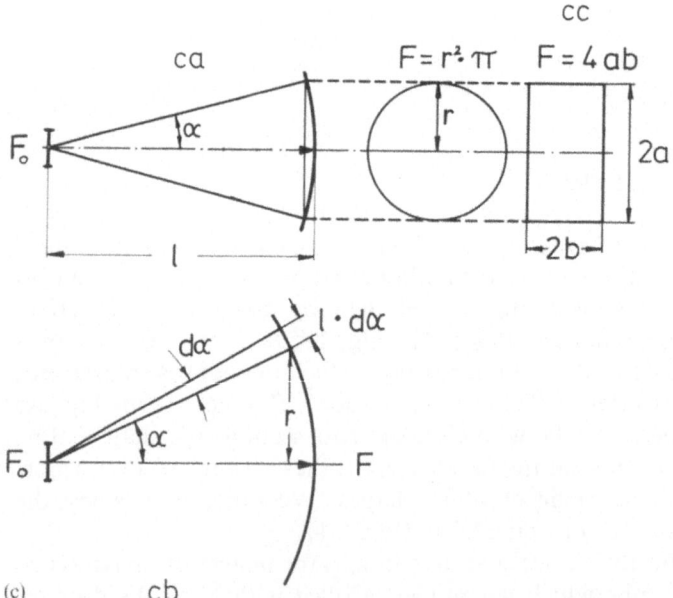

(c)

Fig. 3.1 a–c. Legend see pages 27 and 28

Fig. 3.1 a–c. Parameters determining the optical flux.

(a) Longitudinal section of the light tube in a centred optical system. The tube is assumed to be so narrow that $\alpha \approx \tan \alpha \approx \sin \alpha$; α in unit of arc (rad). Definition of optical flux,

$$OF = \frac{F_1 \cdot F_2 \cdot N^2}{l^2} \tag{3.8}$$

OF optical flux (in $cm^2 \cdot sr$); F area limiting the light tube (in cm^2) (subscripts 1 and 2 indicate the area and the angle having its apex in the area in question); l distance between F_1 and F_2 (in cm); N refractive index of the medium between F_1 and F_2.

For the ratio of an area and the square of length we can make simplifying approximation to obtain solid angles,

$$F_1/l^2 = \omega_1, \ F_2/l^2 = \omega_2.$$

Putting a solid angle into Eq. (3.8) we obtain

$$OF = F_1 \cdot \omega_1 \cdot N^2 = F_2 \cdot \omega_2 \cdot N^2 = r_1^2 \cdot \pi \cdot \omega_1 \cdot N^2 = r_2^2 \cdot \pi \cdot \omega_2 \cdot N^2$$
$$= (r_1 \cdot \sin\alpha_1 \cdot N \cdot \pi)^2 = (r_2 \cdot \sin\alpha_2 \cdot N \cdot \pi)^2 \tag{3.9}$$

ω solid angle (in sr) (For relation between the solid angle and half of the plane angle (α) see Fig. 3.1c); r radius of F (in cm).

Notes:

1. The term ($\omega \cdot N^2$) is called numerical solid aperture. It has the unit of sr.

2. In practice F_1 is a field (or window), then α_1 is an aperture angle; thus $\alpha_1 \equiv u$, and F_2 is an aperture area (or pupil), so that α_2 is half a field angle; thus $\alpha_2 \equiv w/2$, or F_1 is an aperture area, α_1 being half a field angle and F_2 is a field, α_2 being an aperture angle.

(b) Illustration of a solid angle. A cone, having its apex at the centre of a sphere of radius l, intersects the surface in a sphere cap of area F. The unit of measurement of the solid angle (ω) is the steradian, which is a number expressing the ratio of two areas,

$$\omega = (F/l^2) \text{ steradians (sr)}. \tag{3.10}$$

It is clear that the maximum value of a solid angle is 4π.

Notes:

1. This applies to all cones and pyramids, but in this book we shall deal mainly with right-circular cones.

2. For the radius of the sphere, the symbol l is used because the radius corresponds to the length of the light tube (see a).

(c) Relation between a solid angle and plane angles.

(*ca*) Right-circular cone of semi-angle α. For a small solid angle we take $F = \pi r^2$; since $\sin \alpha = r/l$, we have,

$$\omega = F/l^2 = \pi \cdot \sin^2\alpha, \tag{3.11}$$

and the optical flux between F_0 and F is expressed by

$$OF = F_0 \cdot \pi \cdot \sin^2\alpha. \tag{3.12}$$

(*cb*) Extension of Eq. (3.12) to a large plane angle with the help of Lambert's cosine law.

We assume the surface of the cap, cut out from the surface of the sphere by a circular cone, to be subdivided in zones, each having the shape of a concentric ring of breadth ($l \cdot d\alpha$); since $r = l \cdot \sin \alpha$ the area of such a zone is expressed by

$$dF = 2\pi \cdot r \cdot l \cdot d\alpha = 2\pi \cdot l^2 \cdot \sin \alpha \cdot d\alpha \tag{3.13}$$

$dF, d\alpha$ differential of F or α respectively.

Fig. 3.1 a–c (continued)

Each zone accepts a radiant flux from a small flat lighting surface F_0 lying in the centre of the sphere perpendicular to the axis of the cone. According to Lambert's cosine law the area F_0 is reduced to $(F_0 \cdot \cos \alpha)$ for a direction subtending an angle α with the line of the normal to the area F_0; so that the optical flux for the light tube limited at the ends by F_0 and F is expressed by

$$OF = (F_0 \cdot \cos \alpha \cdot dF)/l^2 = F_0 \cdot 2\pi \cdot \cos \alpha \cdot \sin \alpha \cdot d\alpha$$
$$= F_0 \cdot 2\pi \cdot \sin \alpha \cdot d(\sin \alpha). \tag{3.14}$$

By integration over the range $\alpha = 0$ to $\alpha = \alpha$ we obtain the optical flux for that range as

$$OF = F_0 \, 2\pi \int_0^\alpha \sin \alpha \cdot d(\sin \alpha) = F_0 \, \pi \cdot \sin^2 \alpha. \tag{3.15}$$

The comparison of this equation with Eq. (3.12) shows that Eq. (3.12) holds for all solid angles, so that we can remove the restriction previously imposed above.

Notes:

1. If α is in a medium with refractive index N the term $\sin \alpha$ must be replaced by the term $(N \cdot \sin \alpha)$.

2. If α concerns an aperture angle thus $\alpha = u$, Eq. (3.15) can be written in the form

$$OF = F_0 \cdot \pi \cdot NA^2 \tag{3.16}$$

$NA = \sin u$ numerical aperture (in a medium with $N = 1$).

3. For the comparison of the ability of optical systems to transfer light often the concept 'linear conductivity' (Zimmer, 1970) is used. This is simpler than that of optical flux.

It is defined by

$$LC = r \cdot \sin \alpha = r \cdot NA \tag{3.17}$$

LC Linear conductivity; r radius of area F; α half plane angle with apex in F.

The unit of mm is mostly applied to r thus to LC, it may also be mm·rad, if α is measured in radians. The relation between the optical flux and the linear conductivity is

$$OF = (LC \cdot \pi)^2 \cdot 10^{-2} \tag{3.18}$$

where the factor 10^{-2} converts mm^2 to cm^2.

(cc) Rectangular pyramide of base sides $2a$ and $2b$.

The area of the cap is $F = 4a \cdot b$; we symbolise the corresponding half (plane) angles by α_a and α_b, respectively, and obtain, $a = l \cdot \sin \alpha_a$ and $b = l \cdot \sin \alpha_b$, so that the solid angle is

$$\omega = F/l^2 = 4 \sin \alpha_a \cdot \sin \alpha_b \tag{3.19}$$

In practice one of the fields (windows) is taken as one of the areas, the aperture area (pupil) through which the field is illuminated or imaged being taken as the other. For our purpose we need take only one area, the rest of the magnitudes being condensed to a solid angle or numerical solid aperture expressed in terms of the corresponding plane angle. Consequently, the optical flux is determined by the following expression which results from Eqs. (3.9), (3.15) and (3.16),

$$OF \equiv F_w \cdot \pi \cdot NA^2 = (r_w \cdot \pi \cdot NA)^2 = F_p \cdot \pi \cdot \sin^2 (w/2)$$
$$= (r_p \cdot \pi \cdot \sin(w/2))^2 \tag{3.20}$$

OF optical flux (in $cm^2 \cdot sr$)   w   window

F area (in cm^2)   p   pupil

r radius of the area (in cm)

NA effective numerical aperture, where $NA = \sin u$, u being the aperture angle; in this equation the name aperture is applied to marginal rays in any window.

w field angle

It follows that the optical flux is proportional to the area of window and square of numerical aperture with which the window is illuminated or imaged or area of pupil and square of the sine of the half field angle with which the window is seen through the pupil. For small angles the sine can be replaced by the tangent or by the corresponding arc.

Note: If the numerical aperture on the stage-object side of an immersion objective is concerned, it is $NA = N \cdot \sin u$, N being the refractive index of the immersion medium.

Effective Optical Flux in the Microscope. In Figures (2.1 a, c and e) it is shown that the whole light train can conveniently be divided into sectors limited at one end by a field diaphragm and at the other by an aperture diaphragm (or by images of these). The optical flux in the whole microscope is determined by the sector from the stage object to the entrance pupil of the objective (or microscope). This is so because all the other sectors have to be balanced to this. With bright-field illumination the entrance pupil is mostly formed by an image of the condenser-aperture diaphragm, because the numerical aperture of the condenser should be controlled so that it is smaller than the numerical aperture of the objective (Fig. 2.1 e). In this case the sector from the stage object to the entrance pupil has the same volume as that from the stage object to the exit pupil of the condenser which is the image of the condenser-aperture diaphragm, and it is the latter sector which primarily determines the optical flux.

There is a difference between the potential and the effective optical flux; the former applies when the field in the ocular and the aperture area of the objective are fully illuminated, while the latter applies when only part of the field and/or of the aperture area is illuminated

In terms of the microscope we characterise the potential optical flux in the following way,

$$OF = \left(\frac{NA_{Obj} \cdot \pi \cdot FN}{2 \cdot MN_{Obj} \cdot TF} \right)^2 = (NA_{Obj} \cdot \pi \cdot r)^2 \tag{3.21}$$

OF potential optical flux in the microscope (in $cm^2 \cdot sr$)

FN field-of-view number of the ocular (in cm)

MN_{Obj} magnification number of the objective (numeral)

NA_{Obj} potential numerical aperture of the objective (numeral)

TF tube (lens) factor (numeral)

$r = FN/(2 \cdot MN_{Obj} \cdot TF)$ radius of stage-object field (in cm)

(For definition of characteristics see Chap. 6, Sects. 4.1 and 5.)

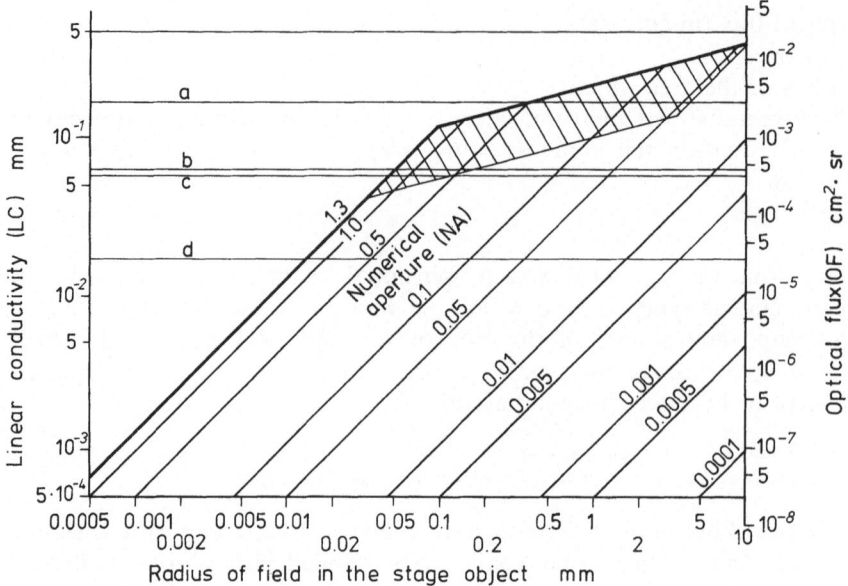

Fig. 3.2. Nomogram relating the optical flux (linear conductivity) to the radius of the stage-object field and the effective numerical aperture of the objective.

Two equations are represented on the same nomogram and the graphs are straight:

Optical flux (right-hand ordinate)	OF	$=(r$	$\cdot\pi\cdot NA)^2$
Linear conductivity (left-hand ordinate)	LC	$=r$	$\cdot\ NA$
	vertical	horizontal	oblique straight lines

All three scales are logarithmic. r radius of the stage-object field; NA effective numerical aperture.

Notes:

1. For the optical flux the radius is given in units of cm, for the linear conductivity in units of mm.

2. The diagram covers all practical values of the parameters involved.

3. Within the hatched area lie the values for standard optics using the full numerical aperture of the objective and the full field in the ocular diaphragm. Values above the hatched area do not occur in practice.

4. Below the hatched area the effective numerical aperture is less than the full or/and the size of the field is less than full.

5. The horizontal straight lines indicate the potential light-transfer parameters for the following:
   a) continuous narrow-band interference strip filter with length of interference layer of 130 mm, typical bandwidth of 15 nm, and typical maximum beam width of 2 mm (Chap. 5, Sect. 3).
   b) continuous narrow-band interference strip filter with length of interference layer of 45 mm, typical bandwidth of 15 nm and typical maximum beam width of 0.75 mm (Chap. 5, Sect. 3).
   c) Grating monochromator (Type M 20) with the monochromator exit having an effective width of 3 mm; bandwidth 15 nm.
   d) Glass-prism monochromator (Type M 4 G II) with the monochromator exit having an effective width of 0.9 mm; bandwidth 15 nm at the central wavelength of 550 nm.

6. For 'typical maximum beam width' of continuous interference filters or relation between width of monochromator exit and bandwidth see Eqs. (5.2) and (5.9)

The same expression is used for determining the effective optical flux, but then we have to measure the value of the diameter of the effective field in the ocular or photometer diaphragm and put this at the place of *FN* and to measure the value of the effective numerical aperture (App. 14). The relations between the parameters involved are shown (Fig. 3.2).

In photometry, the field is kept much smaller than for observation, and, especially for measurements of transmittance (absorbance) or specular reflectance, the effective numerical aperture must be very small (Chap. 11, Sect. 5). It has been shown experimentally that under favourable conditions the smallest optical flux giving a reliable measuring value is about 10^{-8} cm$^2 \cdot$ sr. This could be produced by a circular stage-object field of radius of 0.5 μm with an effective numerical aperture of not less than 1.0.

3.4.2 Influence of Optical Flux on the Size of Areas at Certain Levels in the Optical Train

We now consider the part of the optical train containing the illuminating system. It is clear that this must be capable of providing the maximum optical flux that the imaging system of the microscope is capable of accepting; we can always reduce the illuminating optical flux by reducing the diameter of the luminous-field diaphragm and/or of the condenser-aperture diaphragm, if there is a reason for doing so. It follows that the luminous surface of the lamp must have a certain minimum size and the collector, having a standard diameter, must accept light with a certain minimum solid angle; when the surface or the solid angle is too small, the light cannot fill the full aperture area of the condenser or objective. The proper conditions will be assured by the manufacturer, but any change, and in particular the use of a free-standing lamp and/or collector requires that attention be paid to these relations.

We can illustrate these relations by a numerical example, and we start with the fact that the optical flux is at its greatest when the lowest-power objective is combined with an ocular having the greatest field-of-view number and when there is no tube factor. Let us take an objective and ocular of standard design having the following values; $FN=2$ cm, $MN_{\text{Obj}}=2.5$, $NA_{\text{Obj}}=0.08$, $TF=1$, and put these values into Eq. (3.21); we obtain the quantity of optical flux as 0.01 cm$^2 \cdot$ sr, or of linear conductivity [Eq. (3.17)] as 0.32 mm. Taking a collector of standard design accepting a light cone with a half plane angle of 35° we can express the linear conductivity in the sector between the lamp and the collector as $LC=0.32$ mm $=r \cdot \sin 35° \approx r \cdot 0.6$, whence r (the radius of the luminous surface of the lamp) is about 0.5 mm.

The balancing of optical flux, and hence of linear conductivity impose the following relations between lamp and collector. The size of the luminous surface of the lamp and the light-gathering power of the collector must be such that there is full illumination of aperture areas and fields. If a collector of smaller light-gathering power with the same diameter, hence with larger focal length, is substituted, the light cone will be more acute. Let us say, for example, the half plane angle is less by one-quarter, then the lamp must be replaced by another one having a luminous surface with a diameter broader by about one-third than that of the first lamp. A 'point source' should never be used with the microscope.

Note: Also a laser of standard size does not fit in the normal optical train of the microscope, because it provides a too small angle of divergence, hence a too small optical flux. This is explained as follows:

The diameter of the luminous surface of the laser is about 5 mm, the angle of divergence about 10^{-3} rad. Using these values for the terms in Eq. (3.20) we can take $r_p = 2.5 \cdot 10^{-1}$ cm, $w/2 = 10^{-3}$ rad and obtain the optical flux as about $6 \cdot 10^{-7}$ cm$^2 \cdot$sr. Since, on the other hand, the optical flux of the microscope is between 10^{-4} and 10^{-2} cm$^2 \cdot$sr; it follows that with a laser only an extremely small portion of the aperture area and/or the field is illuminated (see comparison of optical-flux values in Fig. 3.2).

Diameter of the Exit Pupil of the Microscope. Whereas the diameters of diaphragms are easily measured if not supplied by the manufacturer that of the exit pupil of the microscope depends on several factors. Its radius is determined by the aperture angle on the image side and on the focal length, hence magnification number of the ocular. Since this aperture angle is always small, we take

$$\tan u' \approx \sin u' \approx r_{\text{EPM}}/f \tag{3.22}$$

u' aperture angle on the image side
r_{EPM} radius of the exit pupil of the microscope (in mm)
f focal length of the ocular (in mm)
By definition it is true, $f = 250$ mm$/MN_{\text{Ocl}}$ and $\sin u' = NA/MN_{\text{Obj}}$
NA numerical aperture on the stage object side
MN_{Obj} magnification number of the objective
MN_{Ocl} magnification number of the ocular

Consequently we rearrange Eq. (3.22) to give the radius of the exit pupil in mm in terms of numerical aperture, on the stage-object side, and of the magnification numbers of the objective and of the ocular,

$$r_{\text{EPM}} = \frac{NA \cdot 250 \text{ mm}}{MN_{\text{Obj}} \cdot MN_{\text{Ocl}}}. \tag{3.23}$$

Thus we see that the radius of the exit pupil of the microscope is direct proportional to the effective numerical aperture and inversely proportional to the magnifying powers of the objective and of the ocular.

The presence of a tube lens with an imaging factor necessitates the insertion of this factor in the denominator of Eq. (3.23). In general, we can say that the diameter of the exit pupil increases proportionally to the decrease of the overall magnification.

In practice the range of useful magnification lies between $500 \cdot NA$ and $1000 \cdot NA$ [Eq. (2.3)] so that we can replace the denominator of Eq. (3.23) by the term

$$500 \cdot NA \leqq MN_{\text{Obj}} \cdot MN_{\text{Ocl}} \leqq 1000 \cdot NA.$$

This gives us the limits of the useful diameter of the exit pupil of the microscope:

$2\,r = 500$ mm/500 = 1 mm and
$2\,r = 500$ mm/1000 = 0.5 mm.

Diameter of the Photocathode. It is clear that the diameter of the photocathode must be at least as large as the cross section of the light beam in its plane. The photocathode may be placed anywhere behind the primary image, unlike the eye, but it should be placed at a level where the diameter of the light beam is fairly small; to achieve this, it is placed in, or close to the exit pupil of the microscope or, if the diameter of the photometer diaphragm is adequate it is placed close to the photometer diaphragm; it can, of course, be placed in any plane conjugate to either of these levels. The diameter of the light beam at the photocathode is easily measured by inserting a ground-glass screen at that level and examining the illuminated spot under the conditions of adjustment to be used for photometry.

Since the whole area of the photocathode in standard photomultiplier tubes is about 8 mm in diameter, the only danger of this being too small for the light tube occurs when the plane of the photocathode is at a level where the diameter of the photometer diaphragm is wider than this. But the diameter of the exit pupil of the microscope never exceeds 1 mm, so that, if the light sensor were placed at this level, it would be advantageous to insert a diffusing lens in order to spread the light over a larger area of the cathode.

3.4.3 Optical Flux for Diffusely Emitted Light

With diffuse reflection and with fluorescence the aperture angle (the optical flux) of the imaging system influences the radiant flux independently of the aperture angle of the illuminating system, or vice versa. This means that for a given field in the stage object, the radiant flux emanating from it is proportional to the optical flux of both the imaging and the illuminating systems, so that it is advisable to use an objective and a condenser with as high numerical aperture as possible in order to supply a radiant flux as strong as possible to the imaging system.

3.5 Transmittance of the Optical System

The transmittance of the optical system is a measure of the fraction of the light that remains after loss by absorption, reflection, scattering and diffraction; alternatively we may say that it is a measure of the light remaining after interaction between the beam and all the optical elements in the apparatus, apart from the stage object itself. Most of the loss occurs through reflection at lens surfaces, by absorption in the monochromatising device or in polars, and by reflection and absorption at mirrors and beam splitters. We neglect the less

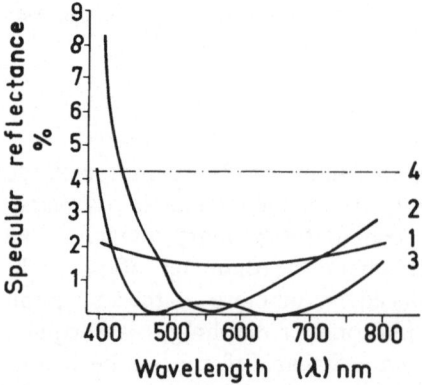

Fig. 3.3. Spectral-reflectance curves of glass surfaces coated for anti-reflection (Anders, 1965). The following table gives the material and arrangement of the layers for the four examples.

Curve No	Manufacturing terms	Materials	Refractive index	Thickness of material either in wavelength units or in nm
1	Glass with single layer	air	1.00	∞
		MgF_2	1.38	$\lambda/4$
		glass	1.52	∞
2	Glass with double layer	air	1.00	∞
		Ce_2O_3	2.15	35.9 nm
		MgF_2	1.38	177 nm
		glass	1.52	∞
3	Glass with threefold layer	air	1.00	∞
		MgF_2	1.38	$\lambda/4$
		Ce_2O_3	2.15	$\lambda/2$
		CeF_3	1.63	$\lambda/4$
		glass	1.52	∞
4	Uncoated glass	air	1.00	∞
		glass	1.52	∞

Notes:

1. The thickness (in wavelength units) concerns optical thickness; mechanical thickness times refractive index of coating material.

2. The anti-reflection effect is due to the extinction of light by perfectly destructive interference of light waves reflected at the upper surface of the layer and the border face glass/layer. This occurs when the path difference of interfering waves is $\lambda/2$. Because one of the waves traverses the layer twice the corresponding optical thickness of the layer is $\lambda/4$

important factors in light loss and express the transmittance (τ_{Os}) by

$$\tau_{Os} = \tau_L \cdot \tau_F \cdot \tau_{Refl.} \cdot \rho_M \tag{3.24}$$

τ_L integral transmittance of all lenses (fraction). For the spectral range in which internal absorption in the lens material can be neglected, this transmittance results from loss of light due to reflection at the lens surfaces. Where the reflectance is ρ (fraction) at a single surface the transmittance of this surface is $(1-\rho)$, and for a number of N surfaces the integral transmittance is $(1-\rho)^N$. Only glass-air surfaces need to be taken into account

τ_F the integral transmittance of filters (fraction) including beam splitters and the monochromatising device. This is the product of the total transmittances (fractions) of the elements involved. (For total transmittance see App. 2.4)

$\tau_{Refl.}$ transmittance of the reflector in the reflected-light microscope (fraction). This results from reflection at, and transmission through the reflector as described in Chapter 6, Section 2

ρ_M integral reflectance of mirrors situated in the optical train (fraction)

The transmittance is increased considerably by the presence of anti-reflection coatings of the highest quality on all glass surfaces (Fig. 3.3), as can be seen from the following rough calculation. For each of the following elements we allow two lenses, thus four surfaces: collector, condenser, objective, ocular. For uncoated glass surfaces we have to allow a loss of 4% (this is, $\rho=0.04$) at each surface, which gives the transmittance as $0.96^{16} \approx 0.5$. When anti-reflection coatings of highest quality are used on all glass surfaces, the light loss at each comes down to about 1% or even less, which results in a transmittance of 0.85, instead of the previous 0.5.

The presence of anti-reflection coatings also improves the image contrast, reduces glare (Chap. 12, Sect. 3.1) and diminishes the depolarisation effect of the glass on polarised light at oblique incidence. The small amount of light that is reflected from these coatings may show a purple or bluish colour, which results from the greater efficiency of the anti-reflection coatings in the middle of the visible spectrum than at the two ends. This colour is sometimes advertised as evidence of the high quality of the coating, but this is misleading since a really high-quality coating would reflect the same low value throughout the visible spectrum and the light would be uncoloured; ideally, there would be no reflected light at all.

In any case, the fewer the inessential glass-air surfaces of any kind in the optical train, the greater is the transmittance and the less the glare likely to be. Of course, all irregularities on the glass surfaces must be avoided (dust, fingerprints, chemical attack, etc.) as these lower the transmittance by scattering the light. Cleanliness of the glass surfaces is essential for the best results, while the microscope should not be kept where the atmosphere is contaminated.

3.6 Interaction of Light with the Stage Object

The interaction consists of absorption, specular and diffuse reflection, fluorescence, interference, refraction, and diffraction.

The following can, as a rule, be measured: absorption expressed in terms of transmittance, absorptance or absorbance; reflection expressed in terms of specular or diffuse reflectance (reflectivity); and fluorescence and interference expressed in terms of relative light intensity. (For definition of terms see App. 2.4.)

These interactions may influence each other and may be disturbed by effects of refraction or diffraction; disturbances can be reduced by appropriate choice and preparation of the stage object.

The loss of light due to interaction with the stage object is outside the control of the observer, who adjusts the optical and electrical modules so as to provide the best instrumental conditions in which to study this interaction by obtaining reliable photometric values.

3.7 Relation Between Light Power and Photometric Measuring Values

The radiant flux reaching the photocathode is transformed into electric current which is transformed in several steps into a quantity, usually voltage, with which the indicating device is able to operate (Fig. 7.4). The relation between the input radiant flux and the output measuring value must be strictly linear. The quotient of measuring value and radiant flux, that is the factor of proportionality of measuring value and radiant flux, is called 'sensitivity of indication' (S_i) (in respect to radiant flux) and expressed as

$$S_i \equiv G/\Phi \tag{3.25}$$

G Measuring value (numeral)
Φ radiant flux (in W)

Notes:
1. An example of determination is described in Appendix 4.

2. The terms G and Φ depend on the wavelength, hence S_i should be specified by the central wavelength if the effective spectral band is narrow, or by the limiting wavelengths, if the band is broad.

General Aspects of Sensitivity. Since there is some confusion about the evaluation of the sensitivity of the measuring device let us consider this matter here and initially define the sensitivity in general as the quotient of a quantity that indicates any reaction and of the quantity that causes the reaction. In our particular case the former quantity is a measuring value (G) or a measuring result that is mathematically related to the measuring value and the latter quantity is any quantity or factor influencing the measuring value. This may be one of the quantities on the right side of Eq. (3.1) as the radiant flux and hence the measuring value depends on these. It depends also on the sensitivity (or response) of the photoelectric assembly so that many factors are involved. We can take any of these as a basis, but this must be specified. In practice, mostly the interaction factor of the stage object is taken; putting x for this factor, the sensitivity can be expressed in four ways:

1. The 'absolute' sensitivity at a given adjustment of measuring equipment is defined by the quotient of G and x. The unit of x is that of the corresponding interaction factor; an example of the latter is the transmittance (T) (in percentage or as a fraction), so that absolute sensitivity is G/T.

2. The 'maximum' (absolute) sensitivity is defined by the quotient of G_{max} and x; G_{max} is the largest possible measuring value that the measuring device is able to produce for a given amount of x, and this occurs at adjustment of the measuring equipment for maximum response.

3. The 'interval' sensitivity at a given adjustment of equipment is defined by the quotient of the two magnitude differences that is by the quotient of (G_1-G_2) and (x_1-x_2).

4. The 'maximum-interval' sensitivity is given by the quotient of the difference of the magnitudes that is by the quotient of $(G_1-G_2)_{max}$ and (x_1-x_2), with the measuring device adjusted for maximum response. Of course, since there will be statistical fluctuations of the measuring values, the mean value must be used for this evaluation, thus $G \equiv \bar{G}$.

3.8 Sensation of Brightness in the Observer

Brightness is a sensation in the eye-brain system of the observer caused by a physical stimulus that can be defined quantitatively. Since our interest is more in the objective phenomenon than in the visual perception, we shall employ the terms of radiation physics rather than those of illumination engineering; the conversion factors from one set of terms to the other are given in Appendix 2.5. Since the human eye is the receptor, we need consider only the wavelength range of its spectral response, namely from about 400 to about 700 nm and having a maximum sensitivity at about 550 nm. It should be noted that CIE tables (CIE; Commission Internationale d'Eclairage) refer to an extreme range of 380 to 760 nm (Fig. A1) but we do not need to use this for our present purpose.

In addition to the light received from the stage object, subjective impression includes also the environment and the psycho-physiological factors of eye adaptation and fatigue. Thus, the subjective impression of the same stimulus varies, not only from one observer to another, but for the same observer at different times; for further information about 'seeing brightness' reference may be made to Gregory (1973) or Schober (1964), where many other literature references are found.

While a photoelectric measuring value is determined by the total radiant flux hitting the photosensor, the impression of brightness in the observer is determined by the density of the radiant flux per unit area. This factor is called either irradiance or else the radiant emittance (radiant excitance), according as an area is assumed to accept or to emit light, both being expressed in the same unit (as W/m^2). Unfortunately different units for area are in use, and so attention must be paid to this for computation with radiometric and photometric magnitudes (see App. 2.5).

The observer senses the radiant emittance of the second (virtual) image (Fig. 2.1e). For computation of the radiant emittance we take the size of the apparent field containing this image and the aperture angle of the light cone, the apex of which lies in this field. We just express the radiant flux emitted by this field after putting $B_\lambda \cdot \Delta\lambda = B$; $OF = F'' \cdot \pi \cdot \sin^2 u''$ and transforming Eq. (3.1) to

$$\Phi = B \cdot F'' \cdot \pi \cdot \sin^2 u'' \cdot \tau_{Os} \cdot IF \tag{3.26}$$

Φ radiant flux (in W)
B radiance of the lamp in a given spectral range $\Delta\lambda$ (in $W \cdot cm^{-2} \cdot sr^{-1}$)
F'' area of apparent field containing the virtual image of the stage object (in cm^2). (The field is taken as being an area of uniform brightness)
u'' aperture angele of light cone emanating from a point in the field and being accepted by the exit pupil of the microscope
τ_{Os} transmittance of optical system (fraction)
IF interaction factor of the stage object (fraction)

Now we compare the parameters by which the optical flux in the sector from the apparent field to the exit pupil of the microscope is determined with those by which the optical flux in the sector from the the stage object to the entrance pupil of the microscope is determined (Fig. 2.1e). Because of the balancing of the optical flux in the sectors the following equation is true,

$$OF = F'' \cdot \pi \cdot \sin^2 u'' = F \cdot \pi \cdot NA^2 \tag{3.27}$$

OF optical flux (in $cm^2 \cdot sr$)
F area of the field in the stage object (in cm^2)
NA effective numerical aperture of the objective on the side of the stage object

Furthermore the following relations can be applied,

$$F'' = F \cdot m^2, \tag{3.28}$$

$$\sin u'' = NA/m, \tag{3.29}$$

where $(m = MN_{Obj} \cdot MN_{Ocl} \cdot TF)$ is the total magnification of the virtual image, i.e. the product of the magnification numbers of the objective (MN_{Obj}) and the ocular (MN_{Ocl}) and, if existing, of the tube factor (TF).

Then the radiant emittance in the apparent field is expressed by the quotient of radiant flux and area of apparent field. This is obtained after rearranging Eq. (3.26) and replacing sin u'' by the right side of Eq. (3.29) so that

$$\frac{\Phi}{F''} = B \cdot \pi \cdot \left(\frac{NA}{m}\right)^2 \cdot \tau_{Os} \cdot IF. \tag{3.30}$$

Equation (3.30) shows that the brightness in the apparent field as seen in the ocular increases directly as the square of the effective numerical aperture and inversely as the square of the magnification.

Notes:
1. For the evaluation of the term Φ/F'' in standard units F'' must be expressed in m^2.

2. It has been found experimentally that the smallest amount of radiant emittance that can just be perceived by the human eye as a very dark light different from total darkness is about $6 \cdot 10^{-13}$ $W \cdot m^{-2}$ for the wavelength of 512 nm. In illuminating engineering this quantity is called the 'absolute threshold of sensitivity for scotopic vision,' i.e. for vision with the dark-adapted eye and given as illuminance in lux (about 10^{-9}lx). In the present case it makes no difference whether the field is assumed to accept or to emit light.

Comparison of Brightness Perceived with and Without the Microscope. We compare the brightness of the second (virtual) image of a stage object with the brightness of the same stage object seen with the naked eye at a distance of 250 mm under the same conditions of illumination. Neglecting absorption of light in the microscope the ratio of brightnesses is

$$\frac{\Phi/F''}{\Phi/F} = \frac{F}{F''} = \left(\frac{r}{r_i}\right)^2 \tag{3.31}$$

Φ/F'' radiant emittance in the second (virtual) image (F'' area of apparent field)
Φ/F radiant emittance in the stage object (F area of field in the stage object)
r radius of the exit pupil of the microscope
r_i radius of the iris of the human eye

Taking the mean radius of the eye iris to be 2.5 mm and that of the exit pupil to lie in the range between 0.5 and 0.25 mm (Chap. 3, Sect. 4.2) the range of the ratio of brightnesses for ordinary microscopic work is expressed by

$$\left(\frac{0.25}{2.5}\right)^2 = 0.01 \leq \left(\frac{r}{r_i}\right)^2 \leq \left(\frac{0.5}{2.5}\right)^2 = 0.04. \tag{3.32}$$

Consequently the brightness perceived through the microscope is 1 to 4% of that perceived with the naked eye.

Derivation of Eq. (3.31). According to Eq. (3.26) and neglecting absorption in the optical system, the brightness of the second image is expressed as

$$\frac{\Phi}{F''} = B \cdot \pi \cdot \sin^2 u'' \cdot IF. \tag{3.33}$$

The brightness of the stage object observed with the naked eye is expressed as

$$\frac{\Phi}{F} = B \cdot \pi \cdot \sin^2 u_i \cdot IF \tag{3.34}$$

Φ radiant flux (in W)
F'' area of apparent field (in cm^2)

F area of field in the stage object (in cm^2)
B radiance of the lamp (in $W \cdot cm^{-2} \cdot sr^{-1}$)
IF interaction factor of stage object
u'' aperture angle of light cone with apex in the centre of the second image and accepted by the exit pupil of the microscope
u_i aperture angle of the light cone with apex in the centre of the stage object and accepted by the naked eye

The angle u is small so that

$$\sin u'' \approx \tan u'' = r/250 \tag{3.35}$$

and

$$\sin u_i \approx \tan u_i = r_i/250 \tag{3.36}$$

r radius of the exit pupil of the microscope (in mm)
r_i radius of the iris of the human eye (in mm)

By replacing $\sin u''$ in Eq. (3.33) and $\sin u_i$ in Eq. (3.34) by the terms on the right side of Eqs. (3.35) and (3.36) and forming the quotient of Eqs. (3.33) and (3.34) we obtain Eq. (3.31).

CHAPTER 4

The Light Source and Its Power Supply

4.1 Introduction

In the present chapter and the three that follow, a description is given of the modules forming the essential of a microscope-photometer system. This division into four is based on the independence of those modules, whereby they need not even be manufactured by the same firm (Fig. 4.1). In the present chapter we discuss the modules by means of which a beam of the correct geometry and power is supplied to the optical system. These comprise the stabilised electric power supply, the lamp, and the collector. For reasons to be discussed in Section 5 a light modulator may be added. The lamp can be either an integral part of the microscope, with holders and carriers, or else it can be free-standing in a housing of its own, this being placed on a bench or on rails so as to bring it into the correct relation to the microscope.

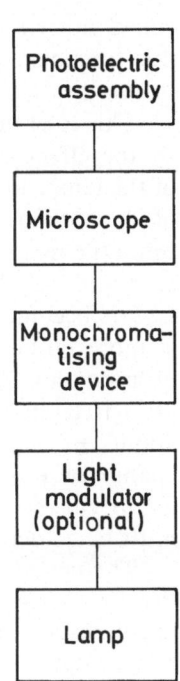

Fig. 4.1. Arrangement of a modular microscope photometer

4.2 Stabilised Electric Power Supply

For photometry the lamp must be supplied with a highly-stabilised direct current, produced by a unit consisting of a transformer, a rectifier and a stabiliser. The efficiency of this stabiliser is expressed by the 'control ratio' and by the amount of the 'residual undulation' or ripple of the secondary voltage (or amperage). The control ratio is the ratio of relative change of the primary (input) quantity (voltage, current or electric power) to the relative change of the secondary (output) quantity, thus

$$\text{control ratio} \equiv \left(\frac{x_{max} - x_{min}}{x_{max}} \right)_i \bigg/ \left(\frac{x_{max} - x_{min}}{x_{max}} \right)_o \qquad (4.1)$$

x   electric quantity (current, voltage, etc.)
max   maximum amount
min   minimum amount
i   input
o   output

The general name for the changes is fluctuation or unsteadiness. This is considered as an AC superimposed on a DC. The amount of fluctuation to be tolerated is determined by the fluctuation that is observed in the actual measuring values. (Rapid statistical changes of the measuring value are often called noise.) The larger the amount of fluctuation of the measuring value due to causes other than the lamp fluctuations, the higher must be the degree of stabilisation of the lamp.

The tolerance for the control ratio of the lamp stabiliser is determined by the effect of fluctuation of the stabilizer-output quantity on the radiance of the lamp. In general, at a wavelength of about 550 nm, the relative amplitude of the radiance fluctuation of an incandescent filament lamp is some three times the relative amplitude of the voltage fluctuation and five times the relative amplitude of the current fluctuation; at a wavelength of about 400 nm these factors become six and ten respectively. This shows that stabilisation of voltage is more efficient than that of current. In most cases a voltage-control ratio of more than 300:1 is required; this means that the fluctuation of input voltage of $\pm 10\%$ must not result in a fluctuation of output voltage by more than about $\pm 0.03\%$. But, voltage stabilisation is effective only when there is no change of resistance in the secondary circuit, as this would produce changes in current; consequently, with voltage stabilisation, the contacts between wires must be soldered, or else firmly fixed by screws.

The need for stability of internal resistance in the lamp practically confines the use of voltage stabilisation to tungsten-filament lamps. On the other hand, for discharge lamps with their rather large variation of internal resistance, current stabilisation is mostly used. For such lamps current stabilisers have a current-control ratio of at least 20:1, i.e. a 10% fluctuation of input current results in at least 0.5% output current fluctuation. In general filament lamps can be stabilised better than discharge lamps. But electric hardware design is improving rapidly, and it is to be expected that soon devices will become available at reasonable cost for achieving true power stabilisation, that is of both voltage and current.

Note: Incandescent-filament lamps can tolerate quite a large amount of ripple (pulsation) of electric power, but discharge lamps cannot do so. For the latter type of lamp the voltage supply must not have a ripple of more than about 0.3% of the operating voltage, i.e. the ripple must not involve a difference of voltage greater than a few millivolts.

In connection with the stabilisation of lamps, two other factors must be taken into account. The first is the tolerance of variation in the nominal input voltage, which is generally about 10%. The second is the variation in output voltage due to changes of temperature. A change of about 0.1% of the voltage for a temperature change of 1°C must be taken into account. The stabiliser should be equipped with a blower in order to keep temperature changes within tolerable limits.

4.3 Various Types of Lamps

4.3.1 Energy Properties

There are two main types of lamps that can be recommended for use in photometry with the microscope — incandescent-filament lamps and lamps where an electric current is discharged through a gas or metal-vapour plasma at high pressure (10 atm or more). The quantities characteristic of both types of lamps are given in Table 4.1.

The power of the lamp can be expressed in three ways:

1. spectral concentration of radiance (in $W \cdot cm^{-2} \cdot sr^{-1} \cdot nm^{-1}$),
2. luminance (in $lm \cdot cm^{-2} \cdot sr^{-1}$) (3),
3. colour temperature (in K) (for the definition of which see App. 2.5).

The first term gives the best specification of the energy characteristics; it indicates the spectral distribution of absolute power, hence is expressed in spectral curves, which can be drawn for a wide range of wavelengths from UV to IR (Fig. 4.2). Luminance is related to visual sensation of light hence can refer only to the visible range. It is mostly given as a single quantity for the whole range, so that it does not give information about the behaviour of the lamp at a certain wavelength or within a small wavelength interval. Colour temperature is the least informative quantity for the subject of this section and need not be discussed in detail.

Incandescent-filament lamps of tungsten emit a continuous spectrum similar to that of a black-body (or full) radiator (de Vos, 1954). Lamps with a flat wire coil are preferable to those with a round wire coil. The luminous element of discharge lamps is a discharge arc in a superpressure atmosphere between two electrodes. When the atmosphere consists of xenon a continuous spectrum with some weak emission bands in the UV, moderate ones in the visible range but strong bands in the IR is emitted. When it consists of mercury, pronounced spectral lines and bands superposed on a continuous background are emitted.[2]

Tungsten lamps and xenon-arc lamps should be used for photometric measurements when continuous scanning through a certain spectral range is required; tungsten lamps for measurements in the visible and near IR range; and xenon lamps for those in the UV and visible range, except in that between about 450 and 500 nm. If the isolation of a spectral line of mercury, e.g. 546 nm is required, a mercury-arc lamp, combined with a monochromatising device can be used. Both mercury and xenon-superpressure lamps are used for excitation of fluorescence, mainly the mercury line at 365 nm.

---

[2] It may be noted that hydrogenium or deuterium lamps are often recommended for macrophotometry in UV light; because of their low radiance, however, they cannot be used for photometry with the microscope.

Table 4.1. Technical data of lamps suited for visual observation and photometry with the microscope

Type			Electric supply		Luminance	Area of lighting surface	Colour temperature	Nominal life time
			Power W	Voltage V	lm·cm^{-2}·sr^{-1}	mm^2	K	h
Tungsten-filament lamps	Flat coil filament		15	6	1000	1.8·1.8	2850	100
	Round coil filament		30	6	1700 (at 5.7 A) 2500 (at 6.1 A)	2.2·2.2	3200 3400	
	Flat coil filament		60	12	1250	3.2·3.2	3050	
			100	12	1600	3.7·4	3150	
	Flat coil filament (Halogen)		100	12	4500	2.3·4.2	3300	300
Discharge lamps	Xenon super-pressure	XBO 75 SSX 75 W/2	75	14 (>50)	40000	0.5·0.5	6100	400
		XBO 150 W/1	150	20 (>75)	15000	0.5·2.2	6100	1200
		CSX 150 W/1	150	20 (>65)				
		XBO 250	250	14 (>60)	26000	0.7·1.7	6100	
		XBO 450 W	450	18 (>70)	35000	0.9·2.4	6300	2000
		CSX 450 W	450					
	Mercury super-pressure	HBO 100 W/2	100	20 (>55)	170000	0.3·0.3	continuum plus lines	200
		CS 100 W/2	100					
		HBO 200 W/2	200	65 (>105)	33000	0.6·2.2		400
		CS 200 W/2	200	57 (>120)				

4.3.2 Other Characteristic Properties

The *areal distribution of luminance or radiance* in the luminous surface depends
on the structure respectively of the filament or of the plasma. The distribution
is more or less irregular because of the empty space between the wires of
the filament, or else because of the increase of concentration of plasma towards
the electrodes of the discharge lamp. The operator must see that the part of
the luminous surface that is imaged in the aperture area of the condenser
is as bright and as uniform as possible. In order to achieve this the lamp
image has to be adjusted, either by displacing the lamp or by displacing the
collector in a direction perpendicular to the optic axis of the illuminating system
(Figs. 8.2 and 8.3).

Uniformity of illumination can be further improved by making use of the
real image of the back surface of the light emitting element that is produced
by a concave mirror fitted in the back of the lamp housing. When this element
is placed at the centre of curvature of the mirror, an image of the back of
its surface is formed in the plane of its front side. The adjustment should
be made to ensure that a surface element of lesser radiance is overlapped by
the image of an element of greater radiance[3] or that the whole image is lying
by the side of the luminous surface (Fig. 8.3). The area of the latter should
not be too large, as this merely causes a waste of electric power through illuminating a portion of the condenser-aperture area outside the effective area.

The stability of the light emission is very important in ensuring precision
in the measuring result, and we have to take into account *instability both in
time and in space.* Instability in time concerns fluctuations in the radiance emitted
by the whole surface, whereas in space it concerns uneven fluctuation at different
surface elements, which is called flicker. Instability in time is due mostly to
variation in the electric supply to the lamp, and it occurs with all types of
lamp. Instability in space is only observed in discharge lamps, and it arises
from plasma migration across the top of the electrodes. It may be caused
by minute variations in internal resistance, or in the power supply, or by mechanical vibrations. Experience shows that the arc in a discharge lamp is more
stable the smaller the distance between the electrodes. Flicker can be kept

---

[3] Overlapping of surface elements with images of elements having the same radiance results in
the destruction of the lamp through local overheating.

Notes to Table 4.1:

1. The voltages in brackets are ignition voltages.

2. The values of luminance are mean values across the lighting area. For comparison: the luminance of the
sun is about $150000 \ \mathrm{lm \cdot cm^{-2} \cdot sr^{-1}}$.

3. Lamps of type BO are made by OSRAM, that of type CS by PHILIPS.

4. The area distribution of luminance is uneven for all lamps, the spatial characteristic of light emission is
uneven for incandescent lamps, almost even for discharge lamps.

5. For incandescent lamps the luminance is continuously controlled for discharge lamps it is not.

6. The nominal life time of incandescent lamps is considerably increased when they are operated with voltage
below the nominal one.

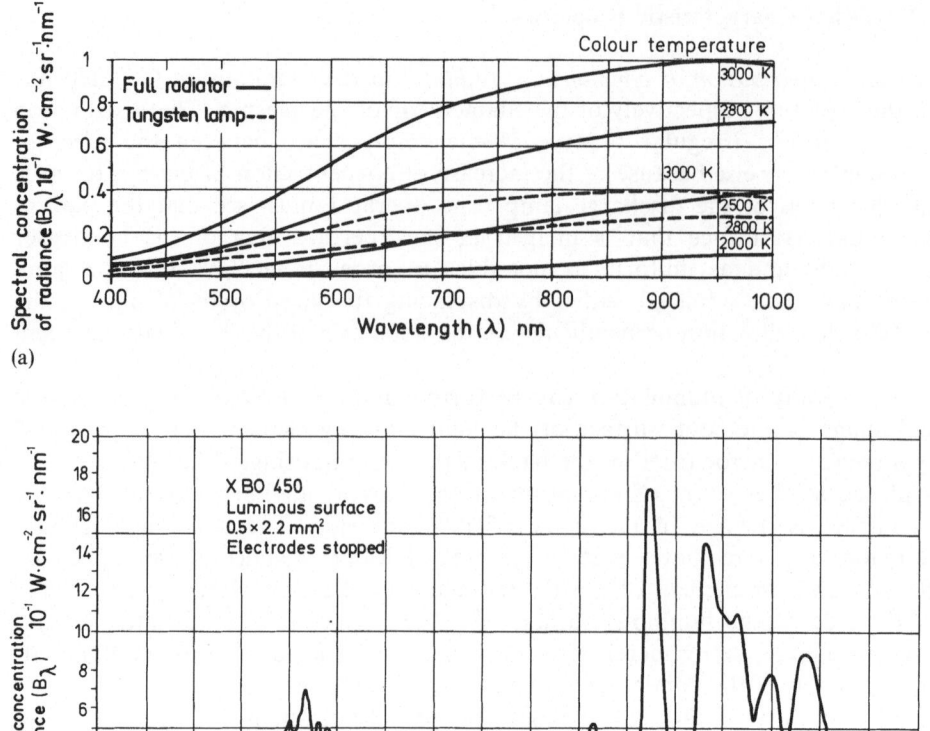

(a)

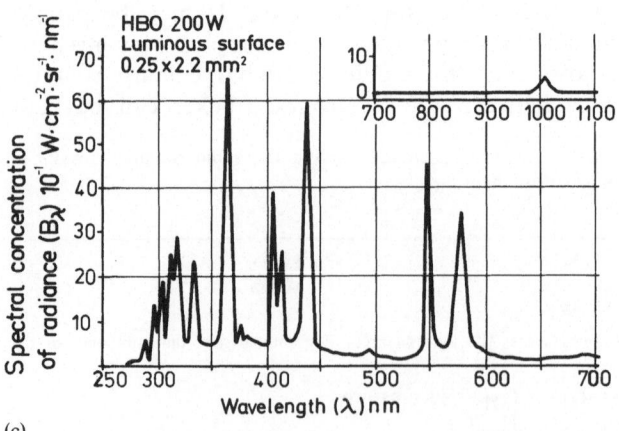

(b)

(c)

Fig. 4.2 a – c. Legend see opposite page

to a minimum by operating a new lamp for about one hour prior to its first use for photometry and then always switching on the power supply at least ten minutes before starting with measurements.

Different individual discharge lamps of the same type may show noticeable differences in stability in space; therefore, they should be examined and selected for stability of this kind. It is clear that flicker will cause the brightest element of the plasma image to move out of the photometric beam. The narrower this beam the more often this will happen.

A sudden change of stability can occur after a long period of work—a discharge lamp does this while in incandescent lamps the filaments can begin to crumble. In any case, a lamp that begins to flicker should be replaced.

Lastly, there is the gradual decrease of radiance with time called *ageing,* this being caused by deposition of evaporated metal on the inside of the bulb. When any darkening of the bulb is observed, the lamp should be changed. An atmosphere of halogen vapour in a tungsten lamp reduces markedly the speed with which darkening proceeds, and such atmospheres are incorporated in lamps of this kind. But halogen lamps are very sensitive to input voltages that are too high, and even a small excess can lead to immediate destruction of the lamp. When the halogen is iodine, the continuous spectral emission between the wavelengths of 500 and 600 nm has a weak ripple due to iodine absorption bands. The amount of ripple is about $\pm 0.2\%$ at 2854 K colour temperature (Sutter, 1973); it decreases with increasing temperature.

The *spatial characteristic of light emission* defines the variation of radiance with direction of emission in space. When this characteristic is irregular, the illuminated field is seen to be unevenly bright, even at strictly Köhler illumination. This effect does not influence photometric measurements made at constant size and position of the measured area, as with object-area scanning, but it does cause errors with varying size and position of the measured area, as with

---

◀ Fig. 4.2a–c. Spectral concentration of radiance of various light sources.
(a) Black-body (full) radiator and tungsten filament. The graphs for the black body show Planck's law of radiation:

$$B_\lambda(\lambda, T) = \frac{c_1}{\pi \cdot \omega_0 \cdot \lambda^5} \cdot \frac{1}{e^{c_2/\lambda \cdot T_c} - 1} \tag{4.2}$$

with the constants $c_1 = 2\pi \cdot h \cdot c^2 = 3{,}7418.10^{-12}$ W·cm^2; $c_2 = (h.c)/k = 1.4388$ cm·$^\circ$K; B_λ spectral concentration of radiance in W·cm^{-3}·sr^{-1} (In this expression the area unit and wavelength unit are condensed to a solid unit in the following way; cm^2·nm $= 10^{-7}$cm^3, but in the figure the unit of cm^2·nm is used.); h Planck's constant (App. 1); c light velocity (App. 1); k Boltzmann's constant ($1.3807 \cdot 10^{-23}$ W·s·K^{-1}); λ wavelength (in cm); ω_0 solid angle unit (1 sr); T_c colour temperature (in K) (For definition see App. 2.5); e base of the natural logarithm.

Note: The ratio of the radiance of a filament to the radiance of the black-body radiator is called emissivity or emission efficiency; the tungsten filament in (a) has an average emissivity of about 0.4 in the visible spectrum for colour temperatures around 3000 K. More accurate emissivity values are given by de Vos (1954).

(b) Superpressure discharge lamp of type XBO 450 W (OSRAM). (c) Superpressure discharge lamp of type HBO 200 W (OSRAM). (The curves of XBO 450 and HBO are published with kind permission of OSRAM GmbH., Munich)

image-area scanning. The most irregular spatial characteristic is that of incandescent-filament lamps (Weber, 1965). This disturbing effect can be reduced by
the insertion of a ground- or etched-glass plate as close as possible to the
front of the lamp (or by light etching of the collector surface that is close
to the lamp). But it is more convenient to place such a plate in a plane where
the illuminating system forms an image of the luminous surface, additional
to that formed in the aperture area of the condenser. But this additional image
must be formed in a plane between the lamp and the luminous-field diaphragm
so that the image of the luminous-field diaphragm is not disturbed by the
ground-glass plate. Of course, with an etched surface light is lost through scattering.

4.3.3 Photometric-Standard Lamps

This section deals with a special type of low-voltage tungsten incandescent
lamp that is designed for direct calibration of macrophotometric equipment
in terms of luminance or quantities derived from luminance. The luminous
element is a tungsten band or filament. Each lamp is supplied by the manufacturer with a chart showing its individually measured spectral-emission properties
at specified conditions of power supply and measurement (Helwig, 1949). Unfortunately the spectral properties are, as a rule, given in photometric quantities;
the manufacturer should be asked for the corresponding radiometric quantities.

   Although this type of lamp is not suited for normal use with the microscope,
it may be beneficial for the calibration of measuring values obtained from
the microscope photometer in terms of radiometric quantities, e.g. radiant flux.
For this purpose, the luminous surface or its image is placed in the optical
train at the level of the stage object so that the optical system of the microscope
images this surface in the photometer diaphragm. The photometric measuring
value from this image is related to the radiant flux striking the photocathode.
The amount of reference radiant flux is calculated with the help of Eq. (3.1);
(B_λ) of the lamp is given by the manufacturer or taken from the literature,
e.g. de Vos (1954); (λ) and $(\Delta\lambda)$ are most conveniently produced with a interference filter situated in front of the photocathode; (OF) is dictated by the effective
portion of the luminous surface (F) and the effective (NA) of the objective
according to Eq. (3.20); (τ_{Os}) is the transmittance of optics between lamp and
photocathode, used for the test, and (IF) is 1.

Note: Such a lamp is especially useful for the calibration of measuring values obtained with
fluorescence (Chap. 10, Sect. 1) if no suitable reference material (Chap. 9, Sect. 3.4) is available.

4.4 The Collector

In this book the term 'collector' is used for both the lens (or lens system)
and the module comprising the lens, the lens mount, a focusing and centring
device for the lens (if present), and an adjustable diaphragm, which is a luminous-

field diaphragm with Köhler illumination. The collector forms an image of the luminous surface of the lamp in the plane of the condenser-aperture diaphragm, or at a conjugate plane, for example, that of the entrance of a monochromator. The lens can be focused so that it forms this image exactly at the proper plane. Alternatively it can be adjusted so that the luminous surface is in its focal plane, thus forming the image first at infinity; an auxiliary (or relay) lens placed between the condenser-aperture diaphragm and the collector is then used to bring the image to the proper place.

For illumination of a stage object to be observed visually, the collector needs to have only a moderate degree of correction. But for photometry a rather high degree of correction is advantageous in order to provide a well-defined image of the luminous surface of the lamp and hence to suppress loss of light due to aberrations. The collector should be capable of being focused so as to allow correction of the position of the lamp image for different wavelengths, unless it has achromatic correction and is preset.

A properly-designed collector permits the illumination of the whole stage-object field and the full condenser-aperture area with a given lamp. Lamps having different sizes of luminous surface must be combined with collectors of different focal length. Given the diameter of the collector, the smaller the size of the luminous surface, the smaller must be the focal length in order to fill the aperture area with the image of the surface. Focal lengths in the range of about 15 to 35 mm are primarily used, while the corresponding half-plane angles of the light cones accepted by the collector are usually in the range of 35° to 17.5°.

For work with visible light and near IR light ordinary glass lenses are suitable, but for UV light special UV-transmitting lenses are required. The highest degree of achromatism in the collector is obtained not with lenses but by means of a concave mirror (or system of mirrors), the luminous surface being placed in, or close to, the focal plane of the mirror. This type of collector can be used for all three spectral ranges.

4.5 Light Modulators

The photocathode can be illuminated either by continuous (direct) light or intermittent (alternating or modulated light). If continuous light is used, then the dark current of the photosensor (Chap. 7, Sect. 2.3), and any stray light entering the optical system from the room outside of the regular light train will produce a background-measuring value superimposed on the value from the stage object. If, however, the light incident on the stage object is modulated with a certain frequency and phase and the operation of electric modules, e.g. the AC-amplifier, is tuned just for this kind of modulation, only the light from the stage object will be accepted by the measuring system, the photocathode dark current and the current due to outside light will be ignored since they are not modulated; thus the background arising from these sources is completely suppressed. The photocurrent noise is also suppressed to some extent, which is especially advantageous with photomultiplier tubes of low to moderate gain and those sensitive for IR light.

The modulation is primarily produced by a mechanical chopper in the shape of a vibrating or rotating diaphragm, and the tuning by a photodiode which senses a phase signal on the chopper for synchronisation of the amplifier system following the photomultiplier. But with such a device the frequency of modulation cannot exceed about 100 Hz; hence, the sampling time (Chap. 7, Sect. 3) cannot fall short of about 0.01 s. If a frequency up to some kHz is required, for example in high-speed area scanning (Chap. 9, Sect. 6.2) or else for detecting very rapid changes in the photometric properties of the stage object, modulation is produced by means of a piezoelectric crystal placed between crossed polars and made to change its birefringence periodically through the action of an alternating electric or magnetic field. This modulation system can be modified so that it measures extremely small effects of optical anisotropy in the stage object (Allen et al., 1966). The chopper method is effective when the light beam has a diameter up to some centimetres, while the piezoelectric one is effective only for light beams having a diameter up to some millimetres.

Note: Photometric devices are now being developed that can pick up measuring signals within time intervals of milli- to micro-seconds (super-speed scanning). This enables tens of thousands of measuring values to be processed within a time interval of a second or less. For this kind of work direct light must be used. Description of this special development is beyond the scope of the present book.

CHAPTER 5

Monochromatising Devices and Filters

5.1 Introduction

Both the material properties to be measured photometrically, the functioning of the optical system and the electrooptical devices depend on the wavelength used. We shall use wavelength to characterise the light, instead of other quantities such as frequency, wave number or photon energy (App. 1), because it is convenient and in common use among microscopists. If necessary, these quantities can be mutually converted. It follows that the measurements must be carried out under specified spectral conditions.

The useful range of commercially-available microscope optics is from about 250 to 1100 nm and this can be divided into three parts: the ultra-violet (UV) from 250 to 400; the visible from 400 to 700, and the near infra-red (IR) from 700 to 1100 nm. A colour filter or monochromatising device either limits a band of given breadth in a given region or else it cuts off all wavelengths above or below a given value.

A broad pass band is sufficient for measuring optical properties which vary only slightly in that spectral range, such as in the densitometry of photographic emulsions. A narrow pass band is necessary in the determination of the concentration (or mass) or organic chemical components in biological specimens, in identifying inorganic substances or, in studying electronic, atomic or molecular structures of microscopic particles by transmitted or reflected light. From a series of such measurements with varying central wavelength in the pass band the complete spectral curve for the property in question can be obtained; from this other properties can be derived. Work in fluorescent light requires the monochromatising device either to block wavelengths shorter than the chosen broad pass band or to block wavelengths longer than the pass band.

A broad band is provided by selectively absorbing filters or else by interference filters having a broad band. The required filter can be selected from the large number of available filters on the market to suit the particular job; their spectral characteristics are given in the catalogues. A narrow band is provided either by a monochromatic interference filter or else by a monochromator. Long- or short-wave blocking is achieved by use of 'chromatic' filters or beam splitters which are called, respectively, 'short-wave pass' and 'long-wave pass.'

In the strictest sense of the prefix 'mono' the wavelength interval would be infinitesimally small, but this would have zero energy, and so it cannot exist. The purer the light in respect to wavelength, the smaller the energy in the beam. In this book the use of the term monochromatic merely indicates that the spectral bandwidth is chosen as small as is permitted by the nature of the experimental conditions. Except where a standard bandwidth is in use, the actual bandwidth should be stated. The experimental conditions comprise both the nature of the substance being measured and the nature of the photometric apparatus.

There are two ways of selecting the pass band; it may be selected by a monochromatising device out of the continuous spectrum furnished by a tungsten-filament or a xenon discharge lamp; alternatively, it may be provided by blocking all except a particular spectral line from a spectral lamp. Of the latter type of lamp only the superpressure mercury lamp is worth considering here since the others have too small a radiance for photoelectric work with small stage objects (small optical flux). Experiments with lasers are being made but, so far, the results have not been satisfactory.

The simplest monochromatising device is a colour filter; this was formerly an absorbing glass filter with a narrow pass band obtained at the cost of a much reduced transmittance. Nowadays interference filters are used and these have a relatively high transmittance for a small width of pass band. For this reason we shall confine our attention here to monochromatic interference filters on the one hand and monochromators on the other.

Homogeneous-Wavelength and Continuous-Spectrum Devices. A 'homogeneous' monochromatising device supplies monochromatic light of fixed central wavelength, and a homogeneous interference filter is used for this. A 'continuous-spectrum' monochromatising device supplies monochromatic light of which the central wavelength can be changed by means of some movement within the typical spectral range of the device. The continuous interference filter is one type, while there are also devices of the prism- and grating-monochromator type.

5.2 Characteristics of Monochromatising Devices

Pass-Band Profile. The first characteristic is the profile of the pass band, which shows the spectral distribution of transmittances within the band. There are three types of profile as shown in Figure 5.1; we take the effective wavelength range of the profile to be that where the transmittance is more than one hundredth of the peak transmittance. Homogeneous filters supply a pass band of constant width and shape. The same do continuous-spectrum filters unless the typical maximum beam width (Eq. 5.2) is exceeded. The pass band of monochromators is variable in width and shape; the ideal shape is that of a triangle (Fig. 5.2).

The profile parameters of the selected pass band determine the spectral purity of the light; this is lowered by 'false light' extended over a wavelength range completely outside the characteristic pass band. False light is usually expressed as the percentage of such light that is mixed up with the light having wavelength within the pass band. The expression is

$$\text{false-light portion} \equiv \frac{I_f}{I(\lambda)} \cdot 100\% \qquad (5.1)$$

I_f intensity of light having wavelengths outside the pass band (in the unit of any radiometric or photometric quantity or as measuring value)

$I(\lambda)$ intensity of light having wavelengths within the pass band (in the same unit as I_f or as measuring value)

λ central wavelength

Since the percentage depends on the spectral position and the width of the pass band, accurate values of percentage can be given only if the width and spectral position is specified. For most cases in practice, it is sufficient to know the average percentage for pass bands in the whole spectral range of operation of the monochromatising device and the pass band is taken as having minimum width (Table 3.1).

Notes:
1. In this book we are not dealing with the measurement of false light because this is rather complicated and special equipment is needed for this procedure.

2. Values of false light are published by the manufacturers of monochromatising devices or will be given on request.

There are three kinds of false light: (1) heterochromatic stray light, (2) fluorescence and (3) harmonic spectral bands. These are described in the following.

1. *Heterochromatic stray light* is light that is irregularly getting through the monochromatising device. This light has an extremely wide spectral band, and so may appear white. It arises from scatter and reflection by or transmission through imperfections or dust on glass faces (Fig. 5.3), cements and metal mounts of the device; it is important that the device be dust-proof and maintained free from dust on the surfaces in the optical train. Holes are unavoidable in the manufacture of continuous-spectrum interference filters, but they may be kept to a minimum by careful selection of the filter for use in the microscope.

The percentage of stray light given by the manufacturer is usually related to the whole area of the dispersing element. It may arise from points regularly distributed across the dispersing element; alternatively it may arise from points or areas accumulated in certain spots. When such spots are imaged, either in the photometric field or in the aperture area, they can well occupy a large portion of what is usually a very small area when photometric measurements are being made with the microscope. It follows that the ratio of stray light to monochromatic light in the photometric beam may be much larger than predicted by the manufacturer. We can imagine the extreme case of a white spot occupying the entire photometric beam with the result that the measuring value would be due only to stray light.

Note: Such a white spot is easily detected in the photometric field or the aperture area if the observer looks at the uniformly illuminated field or aperture area. Of course, the spot should, if possible, be moved outside the photometric beam.

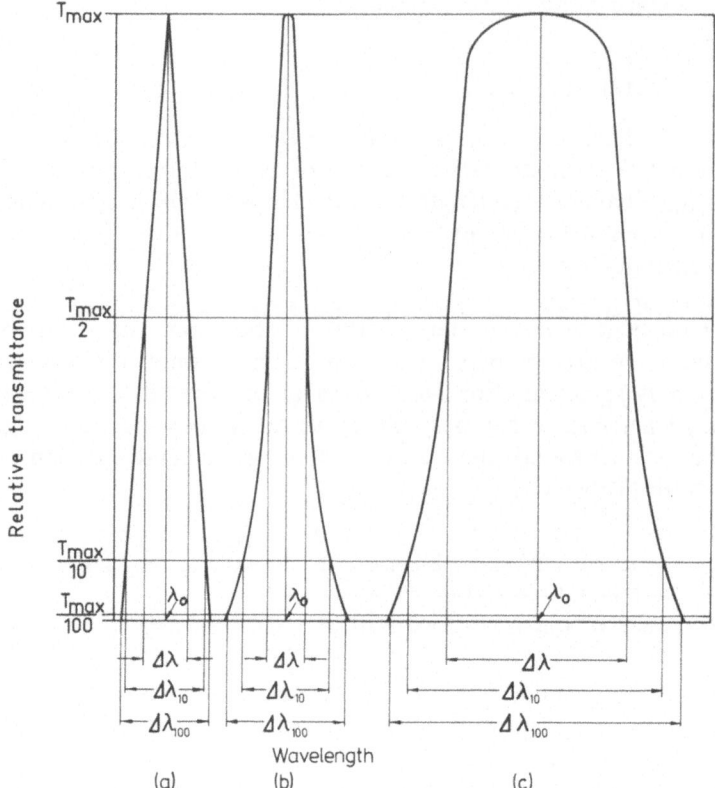

Fig. 5.1a–c. Pass-band profiles of monochromatising devices with denotion of characteristic parameters, schematic.

(a) Profile of a monochromator having the width of the entrance adjusted to that of the exit (*triangle-shaped*) (see Fig. 5.2 for formation of this profile). (b) Profile of a narrow-band interference filter (*spike-shaped*). (c) Profile of a broad-band interference filter (*bell-shaped*).

Notes:

1. The symmetric shape of the flanks is idealized: theoretically, symmetric flanks can occur only if the abscissae are calibrated in units that are proportional to energy, e.g. wave numbers, frequencies (App. 1).

2. Theoretically, for a profile that has no symmetric flanks, a difference must be made between peak wavelength and central wavelength because the transmittance at the latter need not be the maximum. This fact can be neglected for photometry with the microscope; hence, in this book, the terms 'peak' and 'central' will be used of wavelengths (λ_0) synonymously

Notes:

1. When the width of the entrance diaphragm is adjusted to that of the exit diaphragm the monochromator is operated with the highest possible efficiency.

2. When the entrance diaphragm is extremely narrow relative to the exit diaphragm the light entering the monochromator has its absolute minimum; the monochromator is operated with lowest efficiency.

3. When the image of the entrance diaphragm is wider or narrower than the exit diaphragm, the pass-band profile has the shape of a trapezium and the efficiency of the monochromator is not fully used

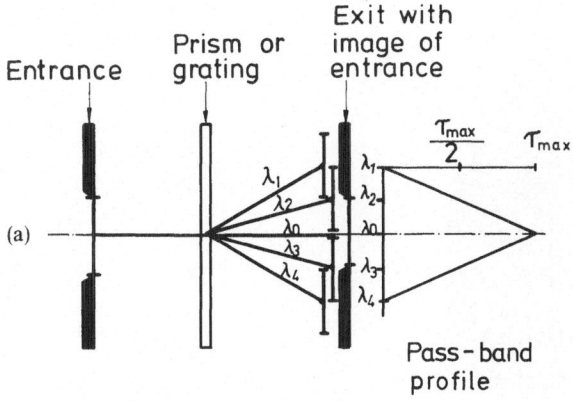

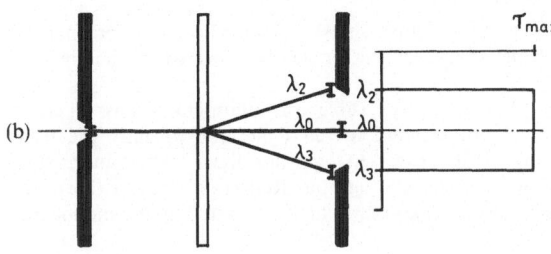

Fig. 5.2 a and b. Formation of the pass-band profile in a monochromator.

(a) Width of entrance adjusted to that of exit. (b) Width of entrance extremely small relative to that of exit. We take the monochromator to be illuminated with 'white' light.

In (a) the width of the entrance diaphragm is adjusted to that of the exit diaphragm. Consequently, the image of the entrance diaphragm, formed with light of wavelength λ_0 for which the monochromator is set, coincides perfectly with the exit diaphragm; hence the amount of light emanating from the exit which we express in terms of transmittance of the monochromator, has its maximum. But, at the same time, the dispersing element supplies light with wavelengths larger or smaller than λ_0. For these wavelengths the image of the entrance diaphragm is displaced in the direction of dispersion. Because the position of the exit diaphragm remains constant, the displacement results in a migration of the image of the entrance diaphragm across the exit diaphragm; thus results in increasing shadowing of the image of the entrance diaphragm, until the image is completely stopped, for which we take the wavelengths λ_1 and λ_4. The shadowing of the image of the entrance diaphragm results in a decrease of transmittance. This decrease is linear with the change of wavelength so that we can take a wavelength λ_2 in the middle of the range λ_0 to λ_1 and another wavelength λ_3 in the middle of the range λ_0 and λ_4 for which the transmittance is half of the maximum; λ_2 and λ_3 are the ends of the nominal bandwidth $(\Delta\lambda)$ by definition.

In (b) the diameter of the entrance diaphragm is taken as being as small as possible, but the exit diaphragm as having the same width as in (a). It is easily seen from this figure that the diameter of the exit widely exceeds that of the entrance for the wavelength λ_0; the image of the entrance diaphragm can migrate across a rather long distance in the exit before it disappears behind the edge of the exit diaphragm. This means that the transmittance of the monochromator is constant within the wavelength interval that corresponds to the distance in the exit (λ_2 to λ_3); but as soon as this interval is exceeded the transmittance is dropped down to zero, and we get a pass-band profile of rectangle shape.

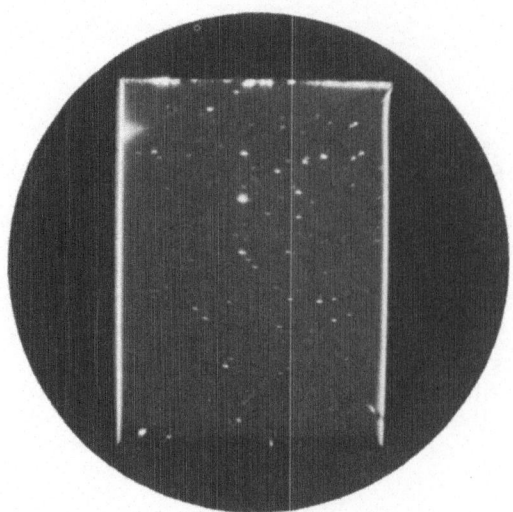

Fig. 5.3. Defects in the monochromator prism. The figure shows a monochromator prism photographed through the microscope, where the image of the prism was formed in the field of the stage object (see also Fig. 5.5 ae).

In order that the white-light spots might be clearly visible, the photograph was taken at a wavelength of 800 nm with a film that did not respond to red light; consequently the back ground is dark. Defects in some part of the prism traversed by white light scatter and reflect this towards the film, as is seen from the spots on the photograph. Reflecting sides of the prism behave in the same way. From such defects heterochromatic stray light is emitted by the monochromator.

Note: This effect can be observed also through the Bertrand lens when the image of the dispersing element is formed in the aperture area of the objective

2. Sometimes glasses and cements in the monochromatising device show a more or less distinct *fluorescence;* its intensity is greater the shorter the wavelength for which the monochromatising device is set. Since fluorescence always extends over a wavelength range considerably wider than that of the pass band isolated by the monochromatising device, it causes an effect similar to stray light. If the monochromatising device is set for a wavelength below about 400 nm fluorescence may be observed as a uniform bluish tint over the field or the aperture area. Monochromatising devices of high quality contain glasses and cements that have no noticeable fluorescence.

3. Grating monochromators and interference filters produce *harmonic spectral bands,* that is bands containing light with wavelengths being integral multiples or submultiples of the wavelengths within the regular band. There are only two cases in which such bands are perceived by the eye; in both cases they result in a purple colour: (1) when a blue harmonic spectral band is superposed on a regular red band or (2) when a red harmonic spectral band is superposed on a regular blue band. Harmonics in other spectral regions cannot be detected directly but must be predicted by the knowledge of the design of the apparatus or become probable by unusual spectral distribution of measuring results.

The *transmittance* of the monochromatising device is one of the magnitudes which dictates the amount of light available for photometric work [see Eqs. (3.24) and (3.1)]. This magnitude is defined by the ratio of light intensity (in any radiometric or photometric unit) reaching the device and having wavelengths within the pass band to light intensity emerging from the device. The term transmittance is, in general, not only applied to devices that work by transmission but also to those that work by reflection, e.g. reflection-grating monochromators or reflecting interference filters.

In addition to their profile, the 'continuous-spectrum' monochromatising devices are characterised by their *dispersing power* and their *dispersing range* (typical spectral range). The dispersing power is expressed as 'linear dispersion.' We take two points defining a distance along the direction of dispersion and derive the quotient;

$$\frac{\text{length (in mm)}}{\text{spectral interval (in nm)}}.$$

This distance can lie either in the exit diaphragm of the monochromator or in the surface of a continuous interference filter. It may be expressed by the inverse term; but in this case it should be called 'reciprocal' linear dispersion. The *typical spectral* range to be used depends on the dispersion of the substance and the design of the apparatus. This range can be defined quantitatively, more or less; usually it is considered to be the range within which the transmittance of the monochromatising device is not less than 10% of its absolute maximum and the false light does not exceed a tolerable amount (e.g. 0.1% or 0.01%).

Note: Values of spectral transmittances and typical spectral range are mostly published without taking into account the effect of any elements blocking false light. These elements diminish the spectral transmittance and restrict the typical spectral range, and so should be taken into consideration.

5.3 Monochromatic Interference Filters

The interference filters described here work by transmission.[4] They consist of a substrate glass onto which have been vaporised alternating layers of low-reflecting transparent dielectric substances and higher-reflecting layers which can be either semi-transparent metal films or else dielectrics of high value of refractive index. The interference principle is not described here, but can be found in textbooks on thin films, e.g. Anders (1965) or brochures supplied by the manufacturers. The substrate glass is sometimes coloured in order to suppress harmonics or else a coloured glass is added in series with the filter.

---

[4] There are also filters which work by reflection. These are of minor importance in microscope photometry, except in microscope fluorometry where they are used to select UV light for producing fluorescence with small wavelengths.

In the simplest case the filter consists of two layers of highly refracting dielectric material or metal with a low-refracting distance layer between, but the number of layers can be some tens; the fewer the layers the smaller the bandwidth and the lower the pass-band profile (spike or narrow-band filter).[5] The more the layers the larger the bandwidth and the higher the peak transmittance but the narrower the base of the profile (bell-shaped or broad-band filter); some typical technical data are given for filters in Table 5.1a and b.

Unfortunately, there is no agreement on the classification of filters from different manufacturers according to bandwidth and tenth-height to half-height widths, etc., and so the same name may be applied to different types. In this book up to a pass-band width of about 15 nm we shall use the term 'narrow-band filter', while for wider bands than this we shall use the term 'broad-band filter'.[5]

The characteristic parameters of the filter are generally expressed with the incident beam collimated and normal to the face of the filter. When the filter is tilted from the perpendicular, or when the beam is otherwise not normal to the face, the pass band broadens somewhat while its centre shifts towards

---

[5] The concept band indicates that the narrowest spectral interval selected by a filter of normal design actually is a band, while that which a monochromator is able to select is as narrow as a spectral line.

Table 5.1a–c. Physical and technical parameters of monochromatising devices (1). (The numbers in brackets refer to notes)

(a) Homogeneous monochromatic interference filters. (Notes for table see p. 61)

Spectral range for which filters are available (one filter for each wavelength)	250 to 1100 nm, other wavelengths on special demand
Filter type and typical bandwidth (in nm)	Narrow band; 0.5 to 15 Broad band; 15 to 80
Transmittance (in %)	10 to 85
Stray light [in % of light within the typical pass band (2)]	<0.01
Standard sizes of filters	Square of $15 \cdot 15 \, mm^2$ to $50 \cdot 50 \, mm^2$ or disc with diameter of 18 mm or 32 mm
Ratio: $\dfrac{\text{Spectral interval at 10\% of peak height of pass-band profile} \quad (\Delta\lambda_{10})}{\text{Bandwidth} \quad (\Delta\lambda)}$	1.5 to 4
Ratio: $\dfrac{\text{Spectral interval at 1\% of peak height of pass-band profile} \quad (\Delta\lambda_{100})}{\text{Bandwidth} \quad (\Delta\lambda)}$	2 to 11

Table 5.1. (b) Continuous-spectrum interference filter. (Notes for table see p. 61)

	Strip filter		Disc filter			
Type of filter	Narrow band in visible light	Broad band in visible light	Broad band in visible and near IR light	Narrow band in visible light	Broad band in visible light	Narrow band in near UV light
Typical spectral range of operation (in nm)	400 to 700		400 to 1000	400 to 700		320 to 420
Typical bandwidth (in nm)	12 to 15	20 to 30	35 to 65	12	17 to 25	15
Transmittance (in %)	10 to 40			10 to 30	No values published	
Stray light [in % of light within the typical pass band (2)]	<0.01			<0.1		No value published
Area occupied by the interference layers (For strip filters: length in mm times height in mm. For disc filters: length of central arc in mm and height in mm of a sector of a ring.) The values in brackets indicate the area of the glass-substrate strip or the diameter of the substrate disc respectively	45·25 a (60·25) 130·25 b (200·25) 120·32 c (140·35)		150·25 d (200·25)	240; 25 e (140)	60; 25 f (100) 120; 25 g (100) 240; 25 h (140)	220; 30 i (140)
			The letters refer to values of linear dispersion listed below			
Ratio: Spectral interval at 10% of peak height of pass-band profile $(\Delta\lambda_{10})$ / Bandwidth $(\Delta\lambda)$	1.5 to 4					
Ratio: Spectral interval at 1% of peak height of pass-band profile $(\Delta\lambda_{100})$ / Bandwidth $(\Delta\lambda)$	2 to 11					
Linear dispersion $(dl/d\lambda)$ (in $\text{mm}\cdot\text{nm}^{-1}$) and typical maximum beam width $(dl/d\lambda)\cdot(\Delta\lambda/3)$ (in mm), the latter is given in brackets (3)	0.15 (0.75) a 0.4 (2.0) b } c	0.15 (1.5) a 8:4 (4.0) b } c	0.25 (6) d	0.8 (3.2) e	0.2 (1.7) f 0.4 (3.3) g 0.8 (6.6) h	2.2 (11) i
			The letters refer to areas of interference layers listed above			

Table 5.1. (c) Prism- and grating-monochromators. (Notes for table see p. 61)

	Dispersing element						
	Prism					Reflection grating	
	Glass		Fused quartz				
	Simple	Double	Simple miniature size	Simple	Double	Miniature size	Normal size
Typical spectral range of operation (in nm)	370 to 2500		200 to 1500	200 to 2500	200 to 1000	200 to 800	200 to 2800
Spectral bandwidth ($\Delta\lambda$) (in nm) for width of exit diaphragm of 1 mm (5, 5a, 6)	15	7.5	75	50	25	80	155 to 70 (6)
Transmittance (in %)	30 to 50	15 to 25	20 to 50	20 to 50	10 to 25	not published	10 to 70
Stray light [in % of light within the typical pass band (2)]	<0.1		<0.6	<0.1		<0.3	<0.1
Ratio: Spectral interval at 10% of peak height of pass-band profile $\dfrac{(\Delta\lambda_{10})}{(\Delta\lambda)}$ Bandwidth	1.8						
Ratio: Spectral interval at 1% of peak height of pass-band profile $\dfrac{(\Delta\lambda_{100})}{(\Delta\lambda)}$ Bandwidth	2						
Linear dispersion ($dl/d\lambda$) (in mm·nm^{-1}) (5, 5a, 6)	0.07	0.14	0.013	0.02	0.04	0.013	0.014 to 0.6

shorter wavelengths. The extent of these changes is determined ultimately by the degree of obliquity of the beam to the face or by the cone angle of the beam. For example with a 10° angle of incidence the central wavelength of 500 nm will shift to 498 nm. Shifts up to this amount can be tolerated, unless the pass-band profile is very pointed and has steep flanks, in which case the obliquely-incident beam can be prevented from passing through at all. A change of temperature will also shift the central wavelength and for an increase of 20°C the shift will amount to about 0.2 nm.

Homogeneous filters have the same characteristics all over the surface, and a standard diameter of 32 mm exists for filters and their holders in the microscope. The continuous-spectrum filters have similar properties, but these are expressed for a typical width of the incident beam, as defined below, or for a smaller width than this. Further they are characterised by linear dispersion by means of which the variation of the central wavelength is related to a distance along the filter. There are two forms of these filters; in strip filters the dispersion is along a straight line, while in circular ring filters it is along the segment of a circle with given radius and given angular extension.

The standard type of continuous filter has a range of about 400 to 700 nm. The dispersing power of a continuous filter is dictated by the steepness of the wedge formed by the interference layers; the flatter the wedge, the greater the dispersion, and also the longer the filter for a given wavelength range. The variation of linear dispersion is approximately proportional to the variation of the length for a given typical spectral range.

Notes for Tables 5.1a–c:

1. The values in the tables are averaged or derived from values given in manufacturers' catalogues, hence the values are of approximate character.

2. The stray light is, as a rule, averaged across the whole surface of the monochromatising element.

3. The linear dispersion of continuous-spectrum disc filters is referred to a length on the central arc of the interference layer. For lengths on the outer or inner margin of the layer the linear dispersion is about 1.3 or respectively 0.7 of that along the central arc.

4. The spectral operation range of grating monochromators is true for gratings that can be interchanged in the same housing; a single grating operates in a range that does not exceed one to two octaves.

5. The bandwidth and linear dispersion of prisms is true for the wavelength of 550 nm. The bandwidths for the wavelengths of 700 nm and 400 nm are related to that for the wavelength of 550 nm by the following ratios;

$$\frac{\varDelta\lambda(700\,\text{nm})}{\varDelta\lambda(550\,\text{nm})} \approx 1.8 \text{ to } 2.2; \qquad \frac{\varDelta\lambda(400\,\text{nm})}{\varDelta\lambda(550\,\text{nm})} \approx 0.3 \text{ to } 0.5.$$

and the corresponding linear dispersions by the ratios,

$$\frac{dl/d\lambda(700\,\text{nm})}{dl/d\lambda(550\,\text{nm})} \approx 0.45 \text{ to } 0.55; \qquad \frac{dl/d\lambda(400\,\text{nm})}{dl/d\lambda(550\,\text{nm})} \approx 2 \text{ to } 3.$$

5a. The values of bandwidth and linear dispersion of gratings are true for the first order of diffraction spectrum.

6. The great range in linear dispersion and bandwidth of gratings as given in the last column of (c) is due to the variation in groove density.

Although, theoretically, the linear dispersion should be the same at all points along the direction of dispersion on the face of the filter, this is not so in practice, and it can vary by amounts up to $\pm 10\%$ of the nominal value. This is due to uneven surfaces in the thin wedges of the vaporised layers. For routine work this can be neglected, but for precise work it should be taken into account by preparing an individual calibration in which the true central wavelengths are plotted against distance along the direction of the dispersion.

Strip filters are manufactured to an effective length of about 40 mm or of about 130 mm, the length of the glass substrate being somewhat greater. With circular filters an effective mean length of about 250 mm is achieved. The dimensions and chief characteristics of commercially-available continuous filters are listed in Table 5.1 b.

There is one parameter of continuous filters that is mostly not given by manufacturers, but which is also listed in this table—the typical maximum beam width. This is the greatest width along the direction of dispersion that is tolerable for the incident beam so as to ensure that the bandwidth will not exceed the typical width. If the width of the incident beam is reduced, the bandwidth is not reduced below the typical width, but it does become larger if the width of the incident beam is too great. The typical beam width is approximately proportional to the product of the linear dispersion and the typical bandwidth, that is $(dl/d\lambda) \cdot \Delta\lambda$. In practice we find the relation to be

$$\text{typical maximum beam width} \equiv \frac{dl}{d\lambda} \cdot \frac{\Delta\lambda}{3} \tag{5.2}$$

$dl/d\lambda$ linear dispersion (in $mm \cdot nm^{-1}$)
$\Delta\lambda$ typical spectral bandwidth (in nm)

In this way we can evaluate the maximum width of incident beam to be tolerated in order to keep the typical size and shape of the pass-band profile. A given percentage excess width of the incident beam produces about the same percentage widening of the band.

We next proceed to study the optical flux of the filters so that we may ascertain the effect of inserting a filter on the size of the field observed (or measured) and on the effective numerical aperture. The optical flux of the filter is expressed by the product of the effective area of the filter and the greatest angle of incidence tolerable if the pass band is to be kept typical in size and shape. We always refer to a light beam with circular cross section and hence obtain from Eq. (3.20)

$$OF = (r \cdot 0.17 \cdot \pi)^2 \tag{5.3}$$

OF optical flux (in $cm^2 \cdot sr$)
r radius of effective area of the filter (in cm)
0.17 rad $\hat{=} 10°$ tolerable angle of incidence

Note: $0.17 \cdot r$ mm is the corresponding linear conductivity (LC).

In the case of a homogeneous filter the value of r is given by the filter radius, while for the continuous filter it is given by half of the typical maximum beam width. Thus we obtain for a homogeneous filter of standard diameter 32 mm the values $OF = 0.73\, cm^2 \cdot sr$ or $LC = 2.7$ mm respectively, where LC is linear conductivity (in mm) and for the 45 mm and 130 mm continuous filters (strip type) (Table 5.1 b) the following values:

Length of effective layer		45 mm	130 mm
Typical maximum beam width	$2r =$	0.75 mm	2 mm
Optical flux	$OF =$	$4 \cdot 10^{-4}\, cm^2 \cdot sr$	$3 \cdot 10^{-3}\, cm^2 \cdot sr$
Linear conductivity	$LC =$	0.06 mm	0.17 mm

If we compare the amount of optical flux of interference filters with the range of optical flux of 10^{-4} to $10^{-2}\, cm^2 \cdot sr$ which is the optical flux supplied by a microscope that is equipped with standard optics (Fig. 3.2) we see that the optical flux of the filters is large enough in order to supply perfect illumination.

It should be pointed out that the homogeneous filter can be placed at any position in the optical system but the continuous filter only in that plane in which the beam width is not greater than the typical beam width of the filter. Since the angle of incidence will not be larger than 10° in any plane other than a plane between the stage object and the objective or condenser, in no case will trouble arise through too large an angle of incidence. The choice of a position in which to place the continuous filter may also be dictated by its design and mode of application. As already mentioned, continuous filters have tiny holes, and they have shifts of colour along the direction of dispersion. Both can be observed when the filter is placed at any plane conjugate to the field, as, for example, at the luminous-field diaphragm. The appearance of tiny bright spots and various colours in the field may be disturbing in visual observation and photomicrography. On the other hand no disturbance may be noted in photometry because the measured field can be chosen so as to avoid all these bright spots.

When a continuous filter is placed at a level conjugate to the aperture area of the objective, the field appears to be uniform since the tiny holes are not seen, and this is dangerous for photometry. For visual observation, however, it is advantageous because the field background appears uniform.

Interference filters are easily spoiled by heat and moisture. They should be carefully cleaned, protected against dust and kept at normal temperature in a dry atmosphere. Finally, we must mention that these filters reflect rather a large amount of light, which may give rise to glare in the image (Chap. 12, Sect. 3.4). These reflections can be suppressed by slightly inclining the filter in the light path so that the normal of the filter forms an angle of about 4° with the axis of the optical system. This should be done with any kind of interference filter.

5.4 Prism- and Grating-Monochromators

5.4.1 Main Properties

Monochromators have a working range between 200 and 3000 nm, but with the microscope only the range between about 250 and 1100 nm can be used. The monochromator is a complex optical system consisting of the dispersing element, lenses (sometimes replaced by mirrors), diaphragms (slits or other shapes), and the mechanism for central-wavelength variation (wavelength drive).

The dispersing material in prism-monochromators is either glass with high transmittance and large dispersion in the visible and IR range, or else fused silica glass with high transmittance in the UV as well but with smaller dispersion in the visible and the UV range. In the grating type of instrument a reflection grating of the 'Echellette type' is used, which has an asymmetric triangular wrinkle profile and in which the monochromatising effect is due to diffraction. This type of grating has a pronounced reflectance maximum at a particular wavelength, called the 'blaze wavelength,' and the blaze is the maximum intensity of light due to specular reflection. For wavelengths larger or smaller than the blaze wavelength the reflectance of the grating (or the transmittance of the monochromator) drops down rapidly, thus the light weakens with the distance of the central wavelength from the blaze wavelength.

A prism operates in a wider spectral range than a grating does because in a prism no harmonics occur. On the other hand, the grating supplies a series of spectra with different orders of diffraction, causing a series of harmonics with wavelengths λ/k ($k = 1$, 2, 3, ...) superposed on the set wavelength (λ). For example, when the typical operating range of the grating begins at 200 nm, the first harmonic can be expected when the setting is at 400 nm.

Harmonics must be suppressed with the help of blocking filters. Some types of monochromators have facilities for convenient interchange, even automatic interchange of filters. Harmonics also can be made ineffective by operating with a lamp that does not emit light in the spectral range of disturbing harmonics and/or using a photocathode that has a spectral response outside the range in which harmonics occur. In practice these three means are applied together and specified by the manufacturer.

The grating is mainly operated with the first-order diffraction spectrum; the use of the second-order one is exceptional, for example for extending the spectral range towards shorter wavelengths. In the visible and IR range the grating, in general, has a greater linear dispersion than a prism and hence a greater optical flux for a given bandwidth [Eq. (5.9)]. The grating allows linear calibration of the wavelength scale, while the calibration in a prism-monochromator is not linear, unless a special mechanism for transformation into linearity is used.

Double monochromators are manufactured both as prism type and as grating type. In these instruments two dispersing elements are aligned in series. They supply a distinctly higher spectral purity of light than do simple monochromators, due to the suppression of false light. Moreover, they have about twice the linear dispersion of a simple monochromator, if the same dispersing element is used twice. For work with the microscope the doubling of the dispersion results in a doubling of the width (in the direction of the dispersion) of the aperture area or of the field that can be illuminated with a given bandwidth. On the other hand, at a given width of aperture area or field to be illuminated, the spectral bandwidth supplied by a double monochromator is half of that supplied by a simple monochromator.

5.4.2 Optical Parameters

The linear dispersion of the instrument is defined as the ratio of the length along the dispersion direction at the exit opening to the difference of central wavelength at the two ends of the length. Because of the change in the linear dispersion of the prism with the change in central wavelength, the linear dispersion has to be specified spectrally. The linear dispersion of the grating monochromator does not depend on the wavelength but on the order of the diffraction spectrum that is being used. The complete mathematical description of the linear dispersion is (Höfert, 1966)

prism-monochromator   grating-monochromator

$$\frac{dl}{d\lambda} = f \cdot 2 \frac{dN(\lambda)}{d\lambda} \cdot \frac{1}{\sqrt{\dfrac{1}{\sin^2(\beta/2)} - N(\lambda)}}, \quad (5.4) \qquad \frac{dl}{d\lambda} = f \cdot \frac{k}{g \cdot \cos\gamma} \quad (5.5)$$

dl	length along the dispersion direction in the plane of the monochromator exit (in mm)
$d\lambda$	change of central wavelength along dl (in nm) ($dl/d\lambda$ linear dispersion)
f	focal length of the collimating lens (in mm)
N	refractive index of the prism material
$dN(\lambda)/d\lambda$	refractive-index dispersion (in nm^{-1}) (The λ in brackets specifies the refractive index for the central wavelength)
β	prism angle
k	order of diffraction
g	distance between two grooves in the grating in nm (grating constant or reciprocal groove density)
γ	angle between the normal to the grating surface and the direction of the diffraction spectrum in consideration

Note: For monochromators of standard technical design the parameters have the following values:

Glass- or Quartz-Prism Type
f lies in the range of about 100 mm to about 500 mm,
$N \approx 1.62$; dN (550 nm)$/d\lambda \approx 10^{-4}$ nm^{-1} (glass)
$N \approx 1.40$; dN (550 nm)$/d\lambda \approx 4 \cdot 10^{-5}$ nm^{-1} (fused quartz)
β lies in the range of 30° to 60°

Grating Type
f is the same as for the prism type
$k = 1$ or 2, according to whether the first or second order is being used

$\cos \gamma \approx 1$, because γ is very small
$g = 833$ nm corresponding to 1 200 grooves·mm^{-1} or $g = 1666$ nm corresponding to 600 grooves·mm^{-1} or $g = 3332$ nm corresponding to 300 grooves·mm^{-1}.

In technical-data sheets for monochromators sometimes values for the spectral resolving power ($\lambda/\Delta\lambda_{min}$) are given. This is the ratio of the central wavelength (λ) to the smallest wavelength difference ($\Delta\lambda_{min}$) that can be separated. For both types of monochromator the equations are

prism   grating

$$\frac{\lambda}{\Delta\lambda_{min}} = b\frac{dN}{d\lambda}, \qquad (5.6) \qquad\qquad \frac{\lambda}{\Delta\lambda_{min}} = k\cdot M \qquad (5.7)$$

b	width of the base of the prism (in nm)
$dN/d\lambda$	refractive-index dispersion of the prism material (in nm^{-1})
M	number of grooves in the grating. This number is expressed by
$M = a/g$; a	width of the grating (in nm)
g	grating constant (in nm)
k	order of diffraction spectrum

Note: The base of a prism of standard design has a width of about 30 mm $= 3\cdot 10^7$ nm, thus, with $dN(550)/d\lambda \approx 10^{-4}$ nm^{-1} the resolving power is about 3000; the separable wavelength difference at $\lambda = 550$ nm is about 0.2 nm. The grating of standard design has a width of about 30 mm $= 3\cdot 10^7$ nm, thus, with $g = 833$ nm the resolving power for the first-order diffraction is 35000; the separable wavelength difference at $\lambda = 550$ nm is about 0.02 nm.

Of course, in order to make use of the spectral resolving power, the exit and entrance of the monochromator must be as narrow as possible; the width of the exit and entrance must not be larger than about 0.01 mm. But a narrow exit does neither allow to illuminate the field or the aperture in the desired size (Fig. 5.5 ac and bc) nor supplies enough light for reliable photometric measurements. Consequently one cannot make use of the spectral resolving power in photometry with the microscope.

5.4.3 Light Transfer Through the Monochromator

We start with the optical path in the monochromator. The instrument has the dispersing element in the centre (Fig. 5.4) with lenses and diaphragms on either side arranged mostly symmetrically to the centre, so that either opening can act as the entrance with the other as the exit. For microscopic work circular, or at least square, openings are better than slits. Complete descriptions of the instruments are to be found in the catalogues of the manufacturers.

The position of the monochromator in the optical train of the microscope is important. Usually it is placed in the illuminating system, and to place it in the image-forming system is exceptional. We shall not discuss this in this book. Proper alignment of the monochromator is possible only with the help of auxiliary lenses properly designed and aligned by the manufacturer of the microscope. In spite of this, the operator should know the principles for the

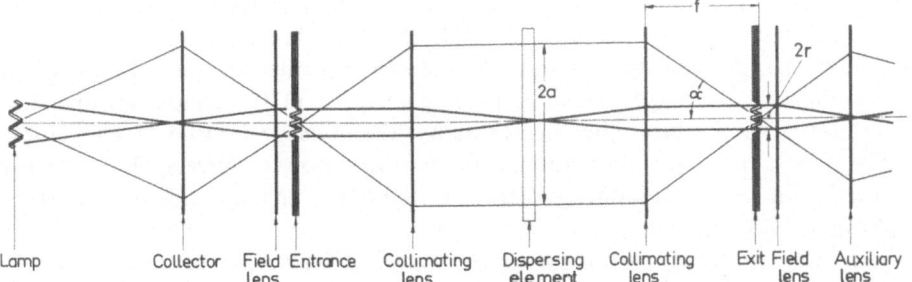

Fig. 5.4. Ray paths in a prism- or grating-monochromator adapted to the ray paths in the illuminating system of the microscope.

Symbols: 2r diameter of the exit along the direction of dispersion; 2a maximum diameter of the light beam in the dispersing element along the direction of dispersion or diameter of a circle inscribed in the projection of the surface of the dispersing element on a plane normal to the optical axis of the monochromator; f focal length of the collimating lens; α angle subtended between the optical axis of the monochromator and a ray joining the centre of the exit to the circle that has diameter 2a.

We take the entrance diaphragm as the surface of the light source and we draw the rays from its centre; by means of a collimating lens they are made parallel at the dispersing element. The second collimating lens brings the rays together at the the centre of the exit diaphragm, so that the entrance is imaged in the exit or vice versa. These rays actually start in the centre of the lamp and are made to intersect in the entrance by means of the collector. After leaving the monochromator they are made to intersect in a plane conjugate either to the condenser-aperture area or the stage-object field by an auxiliary lens (or lens system), so that the exit is imaged in either the aperture area or the field.

The reciprocal rays start parallel at the edges of the entrance, intersect at the dispersing element and become parallel again at the exit. Hence an image of the dispersing element is formed at infinity from either entrance or exit. A field lens placed close to the entrance or exit respectively makes the rays to intersect in planes where the dispersing element is actually to be imaged.

Notes:
1. Monochromators of the simplest technical design contain only one collimating lens, the dispersing element being situated close to it. This arrangement results in moderate light yield and moderate spectral dispersion.

2. In other versions of technical design the effect of lenses is achieved with mirrors. In this figure we have used lenses in order to keep the axis in one line.

3. The auxiliary lens shown in this figure makes the exit of the monochromator conjugate to aperture areas in the microscope and the dispersing element conjugate to fields.

4. The widths of the entrance and exit diaphragm must be adjusted to each other so that the image of the entrance diaphragm exactly fills the exit diaphragm. This adjustment results in the highest possible yield of light for a given wavelength and a given bandwidth because the pass-band profile has the shape of a triangle (Fig. 5.2)

adaption of the monochromator to the ray system of the microscope in order to understand the rules for its control and adjustment.

There are two groups of rays traversing the monochromator; one group consists of rays intersecting in the monochromator entrance and exit and being parallel in the dispersing element; the other consists of rays that are parallel in the entrance and exit but intersect in the dispersing element. The monochromator can be arranged so that the rays of the former group also intersect in the pupils, thus being principal rays and forming an image of the exit in the condenser-aperture diaphragm. Simultaneously, the rays of the latter group intersect in the luminous-field diaphragm, thus being marginal rays and forming an image of the dispersing element in all fields (Fig. 5.5 ae). The monochromator can also be arranged so that an image of the exit is formed in the luminous-field diaphragm and an image of the dispersing element in the condenser-aperture diaphragm (Fig. 5.5 ba). The former arrangement is preferable to the latter.

The most suitable position for the monochromator is between the collector and the luminous-field diaphragm (transmitted light) or the condenser-aperture diaphragm (reflected light).

The lamp is mostly arranged in the simplest way, and this is by merely imaging the luminous surface in the entrance with the help of the collector. On the other hand, a lens must be placed close to the entrance (field lens) in order to image the collector lens in the dispersing element.

Next, we will discuss the geometrical parameters that determine the optical flux of the monochromator linked to the microscope, as this is one of the quantities that determines the efficiency of the monochromator in respect to energy transfer; the other quantity being the transmittance. For the optical flux we consider the areas, limiting the light tube in the monochromator. These areas are circular: one being inscribed in the surface of the dispersing element, the other in the exit slit, because they are respectively conjugate to the stage-object field and the aperture area, which are normally circular. This means that from the whole surface of the dispersing element and from the whole exit only a circular portion is effective for the optical flux, while in macroscopic work the whole surface and the whole exit slit are effective. Actually there is another light tube situated between the dispersing element and the entrance but the size of this is automatically adjusted to that of the former one, so that the latter need not be discussed separately.

In the way that we express the optical flux in the microscope, we express the optical flux in the monochromator by the product of an area and a solid angle; the area is that of the circle inscribed in the exit, the solid angle is that of a straight circular cone having its apex in the centre of the circle and its base in the dispersing element (Fig. 5.4). The expression is

$$OF = \left(r \cdot \frac{a}{f} \cdot \pi \right)^2 \tag{5.8}$$

OF optical flux in the monochromator (in $cm^2 \cdot sr$)
$2r$ diameter of the exit along the direction of dispersion (in cm)
$2a$ maximum diameter of the light beam in the dispersing element along the direction of dispersion, or diameter of a circle inscribed in the projection of the surface of the dispersing element on a plane normal to the optical axis of the monochromator
f focal length of the collimating lens (in the same unit as 2a)
α angle subtended between the optical axis of the monochromator and a ray joining the centre of the exit to the circle that has diameter 2a

(The term a/f is half the plane angle (in rad) of the circular cone; this is true because this angle is small so that $\alpha \approx \tan \alpha$.)

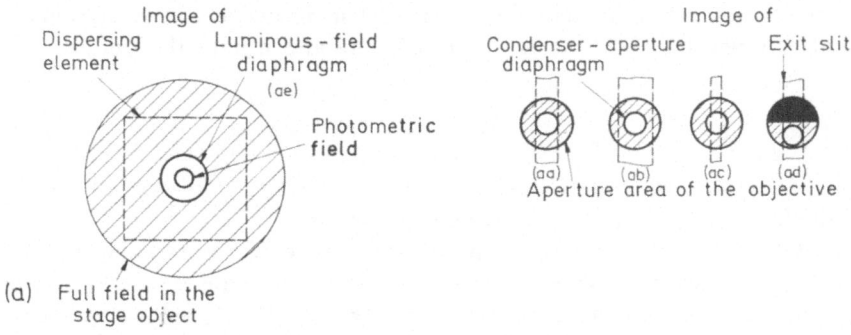

(a) Full field in the stage object

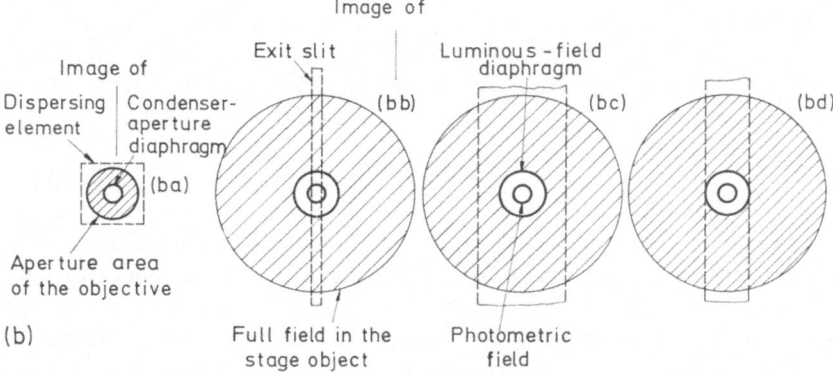

(b)

Fig. 5.5a and b. Adjustment of the image of the exit diaphragm and the dispersing element of the monochromator in level and size. In the figures the exit and entrance diaphragm of the monochromator are taken to be slits, the dispersing element is taken to have a square cross section.

(a) Image of exit slit in the aperture area and of dispersing element in the field. (*aa*) Correct adjustment of the exit slit. (*ab*) Exit slit too wide; the spectral bandwidth is unnecessarily widened. (*ac*) Exit slit too narrow; it restricts the aperture area and hence the effective numerical aperture in one direction, and makes the condenser-aperture diaphragm ineffective in this direction. (*ad*) Correct adjustment of exit slit for operation with a prism reflector in reflected light. (*ae*) Correct adjustment of image of dispersing element (see also Fig. 5.3).

(b) Image of exit slit in the field and of dispersing element in the aperture area [adjustment inverse to that in (a)]. (*ba*) Correct adjustment of the image of the dispersing element. (*bb*) Exit slit too narrow; it restricts the photometric field in one direction, and makes the photometer diaphragm ineffective in this direction. (*bc*) Exit slit too wide; the spectral bandwidth is unnecessarily widened. (*bd*) Correct adjustment of exit slit.

Notes:

1. In order to eliminate superfluous illumination, the exit slit should be narrowed in height or replaced by a circular or a square variable diaphragm.

2. If an extremely narrow spectral bandwidth is required and this restricts the aperture area or the photometric field as in (*ac*) or (*bb*) the lens system imaging the exit must be replaced by one having another focal length and/or another position so that the image of the slit is larger. For this the manufacturer must be consulted

We can express the diameter of the exit ($2r$) in terms of linear dispersion ($dl/d\lambda$) and bandwidth ($\Delta\lambda$); thus as $2r = (dl/d\lambda)\cdot\Delta\lambda$ and obtain the optical flux as

$$OF = \left(\frac{dl}{2\cdot d\lambda}\cdot\Delta\lambda\cdot\frac{a}{f}\cdot\pi\right)^2. \tag{5.9}$$

(For the balancing of units dl must be given in cm.)

Equation (5.9) shows that the potential optical flux of the monochromator linked to the microscope is directly proportional to the square of the linear dispersion and of the length of the shorter edge of the dispersing element, and inversely proportional to the square of the focal length of the collimating lens.

We put the terms on the right side of Eq. (5.9) into Eq. (3.1), which enables us to state that, when the factors other than the bandwidth are constant, the radiant flux in the microscope is proportional to the third order term of the bandwidth, thus

$$\Phi \propto (\Delta\lambda)^3 \tag{5.10}$$

so that the measuring value decreases rapidly with decrease of bandwidth.

Further we must take into account whether the width of the effective light beam at the exit is dictated by the bandwidth or vice versa. Thus, if the exit is conjugate to the aperture area of a given condenser and we wish to illuminate the stage object with a certain aperture angle from a given monochromator, we have no choice of minimum bandwidth; the bandwidth that we obtain is fixed by the width of the exit and the latter is fixed by the width of the aperture area of the condenser because we have to adjust the width of the exit to that of the effective aperture area (Fig. 5.5 aa). If the exit is conjugate to the stage object and we wish to illuminate a certain field in the stage object with a given condenser it is the width of this field to which the exit must be adjusted and by which the bandwidth actually is dictated (Fig. 5.5 ba).

Notes:

1. It is important not to allow any light to enter the illuminating system, other than that from the effective photometric beam at the exit. This is achieved by means of diaphragms at the exit and the entrance, the width of which is adjusted to the effective area of the illuminating system in the planes where they are imaged. Correctly adjusted diaphragms not only suppress stray light, but also correctly limit the bandwidth to what is required by the diameter of the photometric beam. If the diaphragms are too wide at the exit and the entrance, the result is a broadening of the bandwidth without any gain in optical flux.

2. For a given lamp the radiance is directly proportional to the bandwidth [Eq. (3.7)]. If it is desired to increase the radiance with the same lamp in order to increase the measuring value we can only increase the bandwidth by widening the monochromator exit. In this case it must be tolerated that the image of the exit in the condenser-aperture area is wider than the aperture diaphragm.

3. If, on the other hand, a bandwidth is desired that is smaller than that which corresponds to the adjustment of the width of the monochromator exit to the effective conjugate area in the microscope illuminating system, we can adjust the height of the monochromator exit to the height of the condenser-aperture area or photometric field respectively, while narrowing down

the width of the exit to the desired bandwidth. In this case only a strip-shaped area in the condenser-aperture area or in the photometric field respectively is illuminated, the rest is shadowed, unless the diameter of the condenser-aperture diaphragm or of the photometer diaphragm respectively is correspondingly reduced.

4. If the effective aperture area or the photometric field is determined by the image of a slit-shaped monochromator exit the optical flux varies linearly with the bandwidth, while the radiant flux varies as the square of the bandwidth.

5. Gratings produce a considerable amount of polarised light. The polarisation of this light may interact with the polarisation effect of mirrors or beam splitters and/or of a polariser situated in the optical train so that the radiant flux is diminished.

Two examples will show the importance of Eq. (5.9). Let us calculate the optical flux of two monochromators; one having a glass prism, and the other having a grating:

Grating (Type M 20)		Glass prism (Type M4GII)
26 mm	Width of dispersing element $(2a)$	26 mm
330 mm	Focal length of collimating lens (f)	330 mm
0.2 mm·nm^{-1} (1st order diffraction spectrum)	Linear dispersion $(dl/d\lambda)$	0.06 mm·nm^{-1} (at 550 nm)
	Given a bandwidth of 15 nm the following values result from Eq. (5.9)	
3 mm	Diameter of illuminated area in the monochromator exit $(2 \cdot r)$	0.9 mm
$3.4 \cdot 10^{-4}$ cm^2·sr	Optical flux (OF) (plotted in Fig. 3.2 as ordinate values)	$3.1 \cdot 10^{-5}$ cm^2·sr

In general we can state the following:

1. At a bandwidth as great as that given by continuous interference filters, the optical flux of simple monochromators is less than that of filters.

2. In order to obtain from a monochromator an optical flux comparable with that from a continuous interference filter, the monochromator exit must have a width of some millimetres, which could give a bandwidth of some tens of nanometres, whereas that given by the filter is only about 15 nm.

3. The potential optical flux of the monochromator does not allow the illumination of the whole field and/or operation with the full numerical aperture of the microscope optics; hence some shadowing in the field and/or in the aperture area must be tolerated in visual observation with bright-field illumination. Photometric work, on the other hand, is in general restricted to small fields and/or small aperture areas and these can be illuminated by a monochromator. For kinds of illumination other than bright field, the use of a monochromator is exceptional and will not be discussed in this book.

5.5 Filters Other Than Monochromatising

5.5.1 Neutral-Density Filters

These filters are used for the control of light power at constant spectral composition of the light, and some uses are: for protecting the stage object and/or photosensor from excess of illumination; checking the linearity of radiant flux and measuring value (Chap. 7, Sect. 3); balancing of light in two separate optical trains (Fig. 9.1c).

The action of neutral-density filters is either absorption or interference. Absorbing filters usually have some small selective spectral effect, while interference filters are quite neutral throughout their range.

By means of a set of homogeneous neutral-density filters the brightness can be controlled in a stepwise manner. Continuous neutral-density filters provide continuous control and are manufactured either as strips or as discs; they can be used for automatic adjustment of brightness or of a measuring value, in which case they are moved by an electronically controlled motor to the position corresponding to the required brightness. In this way, for example, the measuring value for the reference material can be kept automatically at a constant level. In order to achieve uniform illumination of the plane to which the filter is conjugate two continuous filters must be moved across the light beam in opposite directions.

Neutral-density filters are manufactured with a transmittance down to about 0.001 (0.1%), which corresponds to an absorbance up to about 3.

5.5.2 Heat-Protection Filters

These filters should be put in the light train when there is a danger of overheating of any optical module or else the stage object; the softening of cements, the production of tension in glasses and of physico-chemical reactions in the stage object are examples of troubles produced by overheating. Absorbing filters and interference filters are often combined so that the interference layers are deposited on to heat-absorbing glass. For photometry, filters should be selected that do not have distinct spectral absorption maxima within the working spectral range. Heat-absorbing filters can be destroyed by tensions arising from large local differences of heating in the surface of the filter during operation. It is necessary, therefore, to place such filters at a place where the impinging light pencil has a large diameter, as for example, near the lamp collector lens on the side towards the microscope but before the light has passed through the luminous-field diaphragm.

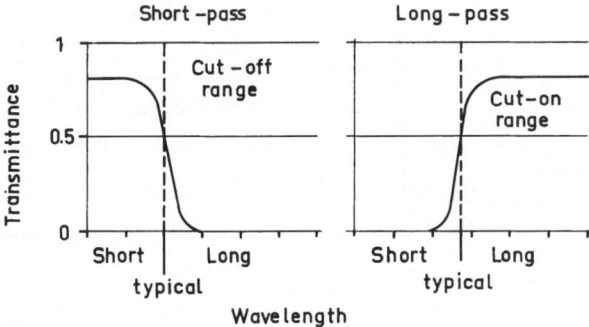

Fig. 5.6. Transmission characteristics of cut-off and cut-on filters

5.5.3 Short-Wave and Long-Wave Pass Filters

These filters have transmission in a relatively wide spectral range but shorter or longer respectively than a specified wavelength, above or below which the transmittance drops down to zero. The typical wavelength (Fig. 5.6) is that at which 50% of the incident light is transmitted. By convention we approach from the short-wave end, and so the name 'cut off' is applied to a filter transmitting the short waves, while 'cut on' is applied to a filter transmitting the long waves. These filters are of the absorbing type, or the interference type or a combination of both; they have three main uses:

1. For blocking harmonic-peak transmission of interference filters or gratings.
2. For selecting light with wavelengths exciting fluorescence in the stage object, these wavelengths mostly being below the middle of the visible spectrum.
3. For blocking just the light that has been selected by the excitation filter and allowing only fluorescent light to pass, in which case they act as barrier filters.

The excitation filter is of the cut-off kind, the barrier filter is of the cut-on kind. For work with transmitted light the excitation filter is placed between lamp and condenser, the barrier filter between the objective and the photocathode; for work with reflected light the excitation filter is placed between lamp and reflector, the barrier filter between reflector and photocathode. Recently, for reflected fluorescent light a type of filter has been developed which acts at the same time as excitation filter, barrier filter, and reflector. It is called chromatic beam splitter or chromatic reflector (Fig. 6.4).

5.6 Polarising Devices

For the measurement of birefringence or of bireflectance linearly polarised light must be used. This can be produced by a calcite polarising-prism or else by a polarising filter, called polar for short in this book; we can proceed to compare the two types.

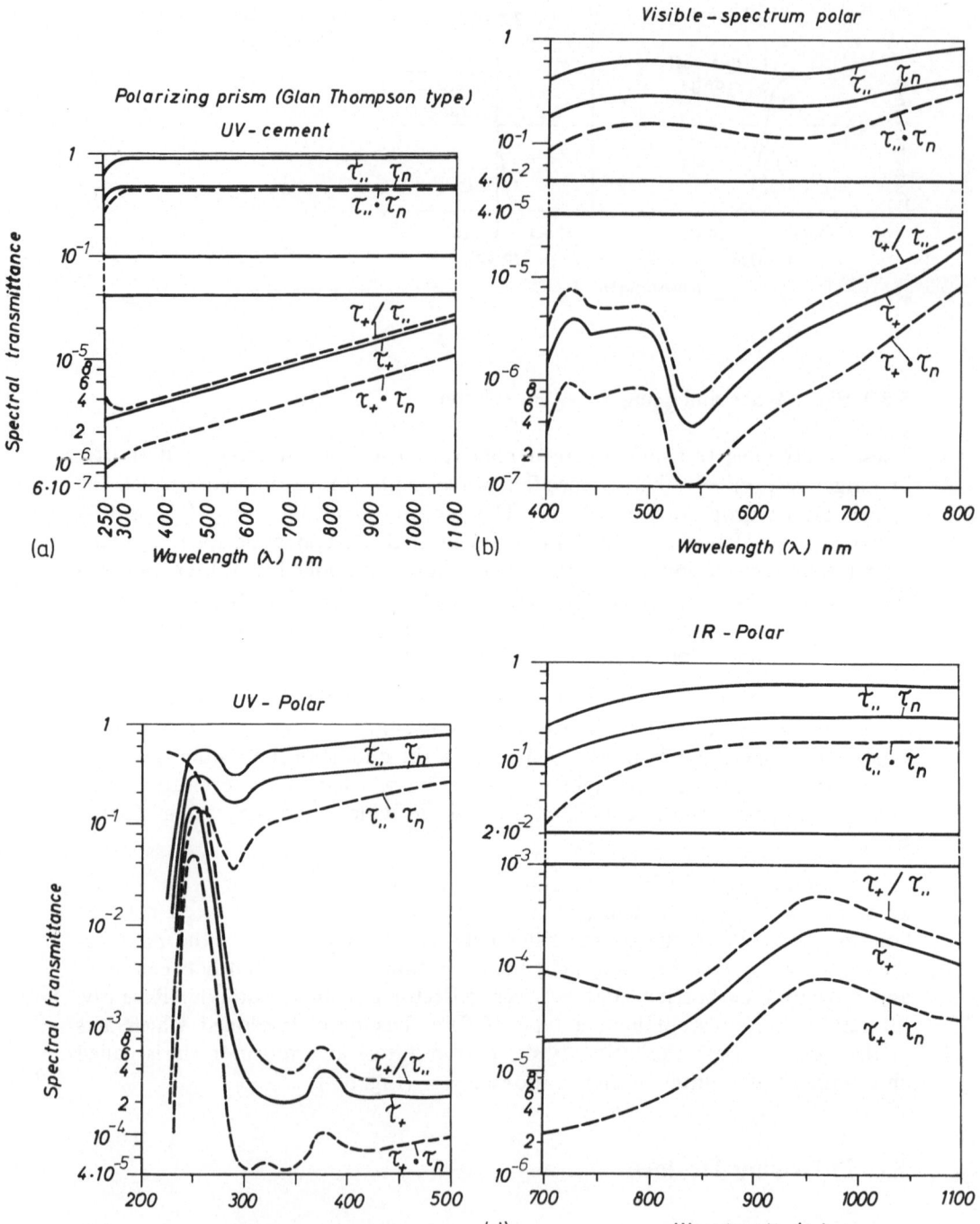

Fig. 5.7a−d. Legend see opposite page

The polarising effect of calcite prisms is based on the splitting of a beam of natural light into two linearly polarised beams having vibration directions at right angles to each other; one of these beams is reflected out of the light train by a layer of cement, while the other is transmitted. The polarising effect of a polar is based on the absorption of one of the two vibrations produced by the anisotropy of the material, which is composed of stained organic macromolecules, (mostly of polyvinylalcohol) linearly orientated by stretching of a foil. In order to avoid disturbance by aging or humidity the foil is cemented between two strain-free glass plates of highest optical quality.

There are polars of different properties specially designed for the visible or the UV or the near IR range, while calcite prisms can be used for the visible and IR range, or, if specially cemented for all three ranges. The transmission properties of polars suited for use with the microscope and those of a calcite prism for all ranges are shown in Figure 5.7. Some other essential properties of polarising devices are given in the following list.

In general, we can state that, in the visible range, a properly designed polar is superior to the prism; in the near IR range suitable polars can be used without much loss of efficacy; in the UV range a prism should be used.

◀ Fig. 5.7a–d. Typical spectral-transmission parameters of polarising devices.

(a) Polarising prism (Glan Thompson type) with UV cement; (b) visible-spectrum polar; (c) UV polar; (d) IR polar.

Full curves: Measured parameters; *dashed curves:* parameters calculated from the measured ones.

Measured parameters: τ_n transmittance of the polarising device with natural (non-polarised) light; $\tau_\parallel$ transmittance of the polarising device with polarised light emanating from a polarising device of the same kind, the vibration directions of both devices being parallel; τ_+ transmittance of the polarising device with polarised light emanating from a polarising device of the same kind, the vibration directions of both devices being crossed.

Parameters calculated from the above: $\tau_n \cdot \tau_\parallel$ integral transmittance of two polarising devices of the same kind placed in series at parallel position and supplied with natural light; $\tau_n \cdot \tau_+$ integral transmittance of two polarising devices of the same kind placed in series at crossed position and supplied with natural light; $\tau_+/\tau_\parallel$ the transmittance ratio $(\tau_n \cdot \tau_+)/(\tau_n \cdot \tau_\parallel)$. Experimentally this ratio is obtained by rotating from the crossed to the parallel position. (By some people the reciprocal of this ratio is used and is called extinction factor)

Notes:

1. The transmittances are *total* transmittances.

2. The efficiency of polarising devices is also evaluated by the 'polarisation effect,' which is defined by $[(\tau_\parallel - \tau_+)/\tau_\parallel] \cdot$ 100%, or the 'degree of polarisation,' which is defined by $[(\tau_\parallel' - \tau_+')/(\tau_\parallel' + \tau_+')] \cdot$ 100%.

The 'polarisation effect' applies to two polars that need not produce perfect polarisation but must be of the same kind. On the other hand, the 'degree of polarisation' applies to a single polarising device arranged in series with another polarising device that produces perfect polarisation. This condition is indicated by the primes.

3. When we calculate the polarisation effect of the materials to which the figures are related we find values larger than 99.998% for both the calcite prism and the visible-spectrum polar in the whole useful spectral range; this is practically perfect polarisation. For the IR and UV polars the values go down to 99.95%, which still can be designated as very good, except, in the UV polar in the range of the transmission band (around about 250 nm) they go down to about 60%

	Calcite prism	Polar
Transmission properties for non-polarised light, qualitatively	High transmittance, spectrally neutral	Moderate transmittance, usually spectrally neutral but spectral variation in transmittance possible
Efficacy in respect to polarisation	High in the UV, visible and near IR range	High in the visible and near IR range, moderate in the UV range; special design for each of these ranges necessary
Shape	Prism with width up to about 20 mm and height up to about 40 mm, or cube with edge length of about 30 mm. (Large prisms are extremely expensive!)	Circular or square plate with thickness up to several mm and diameter up to some tens of cm
Tolerable angle of incidence of light in order to maintain the efficacy in polarisation	Up to 4°.	Up to 10°
Suitability for insertion in the light train of the microscope	Bad	Good
Sensitivity to heat and moisture	Moderately resistant; destroyed by excess humidity and a temperature higher than 70 °C	

Two polarising devices are used in the microscope, the first being the polariser and the second the analyser. In transmitted light the polariser is placed just in front of the condenser, while the analyser is placed between the objective and the ocular, in a plane where the marginal rays are parallel. In reflected light the polariser is placed just in front (i.e. on the lamp side) of the reflector, while the analyser is placed in the same position as in transmitted light. The polarising device, which may be rotatable, is contained in a holder, which enables it to be switched in or out of the light train, either on a slide or

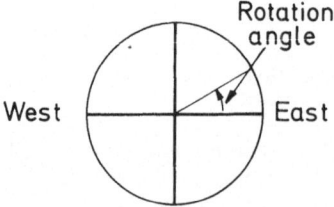

Fig. 5.8. Standard directions and rotation angles of polarising devices in the microscope (after German Standard DIN 58879).
    Zero direction of polariser East-West. When the analyser is in crossed position to the polariser, it is said to be in 90° position. Rotation angles are measured positively in counterclockwise direction

in a swing-out fitting. Instructions for the proper use of polarising devices are given in textbooks on polarising microscopy. In photometry the analyser is mostly in the OUT position, except for work with polarised-light interference equipment (Chap. 10, Sect. 2).

In some makes of older microscopes the N-S direction is that of the permitted vibration in a polariser but, in 1972 most manufacturers agreed to the standard recommendations shown in Figure 5.8 including the E-W vibration (German Standard DIN 58879) and this is being widely followed nowadays.

Note: The vibration plane of light (electromagnetic radiation) is the plane containg the electric-field vector, the polarisation plane is perpendicular to the vibration plane; it contains the magnetic-field vector.

CHAPTER 6

The Basic Microscope Assembly

Note: This chapter deals only with the parts that are common to all research microscopes. Reference should be made to the contents for other parts of the microscope system.

6.1 Condensers

6.1.1 Requirements

We can state that the two essential requirements of a condenser are:

1. It must have a numerical aperture at least equal to that of the objective with which it is to be used.

2. It must be designed to illuminate the whole field in the stage object that is to be studied.

A single condenser, may, of course, be used over a fairly wide range of numerical aperture and field diameter, but this can be done only at the cost of a falling off in optical correction and an increase of glare the further the working conditions depart from those for which it was designed. Some characteristic parameters are listed in Table 6.1.

In photometry with the microscope the need is for high correction, and so a series of condensers with achromatic-aplanatic correction is required for the best performance. On the other hand, both the diameter of the aperture diaphragm and of the luminous-field diaphragm are reduced considerably. The best correction is obtained when the condenser is another objective with magnifi-

Table 6.1. Characteristic parameters for condensers of standard design

Focal length f	Nominal numerical aperture NA	Diameter of aperture area	Diameter of the image on the stage of a luminous-field dia-phragm of 20 mm actual diameter
30 mm	0.3	18 mm	3.75 mm
15 mm	0.6	18 mm	1.88 mm
10 mm	0.9	18 mm	1.25 mm
7 mm	1.4	20 mm	0.9 mm

Notes:

1. The diameter of the condenser-aperture area is calculated from the formula $2 \cdot f \cdot NA$.

2. For a distance of 160 nm between the luminous-field diaphragm and the condenser-aperture diaphragm the values in the last column are given by the formula $20 \cdot f/160 = f/8$.

cation number and numerical aperture just a step smaller than those of the objective to be used for measurement.

When the numerical aperture of the condenser is larger than about 0.3, a correction for thickness and refractive index of glass has to be made; this will be for the thickness and refractive index of the glass slide in transmitted light, for which the standard values are: thickness (1.1 ± 0.1) mm, refractive index (for $\lambda = 546$ nm) 1.52 ± 0.01 (German Standard DIN 58 884).

If an objective is used as the condenser the distance between the luminous-field diaphragm and the stage object has to be the same as that between the stage object and the nominal plane of the primary image. In Section 6 there is an explanation of the difference between the nominal and the actual tube length; this applies also to an objective used as a condenser, and a correcting lens similar to the tube lens may be required.

Quick interchange of condensers is required. A set of condensers may be mounted on a revolving piece, just as objectives are. One or more of the upper lenses in a condenser may be removed or exchanged for others.

In some designs focusing and centring devices are fitted to the condenser lens, the aperture diaphragm and the luminous-field diaphragm being precentred and fixed. In others the condenser lens is precentred and focusable, but the aperture diaphragm and the luminous-field diaphragm is moveable in its own plane. The diaphragms are usually iris diaphragms, but these may be replaced by sets of holes, which enable a very small aperture area and a very small field to be illuminated. This is important for photometric work.

Ring-aperture diaphragms are supplied for work with phase-contrast and dark-field illumination. Slit diaphragms are used for some kinds of interference microscopy. For differential-interference contrast a birefringent beam-splitting prism can be placed in the aperture area of the condenser. These elements are easily interchanged by means of slides or a revolving plate.

6.1.2 Special Designs

It is convenient to group here some notes on special features.

1. Polarised-light condensers. For use in polarized light the condenser must be strain-free, like objectives.

2. Reflected-light condensers. In reflected light the objective of the microscope is necessarily also the last stage of the condensing system. Thus, the condenser is automatically adapted to the objective, including the polarised-light requirements and the effect of immersion.

3. UV-light condenser. This has to be made of special glass and they are corrected to the same degree as are objectives for this purpose (Sect. 4.4).

4. Immersion condensers. A condenser with a nominal numerical aperture greater than 0.9 must be oiled on to the base of the glass slide carrying the specimen. This kind of condenser is not bloomed on its upper surface; thus, if it is used in air its correction is disturbed, and unwanted reflections appear, producing glare.

5. Zoom or pancratic condensers. These may or may not have also a lens exchange device combined with a lens system that, at constant focusing position, enables continuous variation of the diameter of the luminous field in the stage object (zoom lens). At the same time there is an inverse continuous variation of the diameter of the condenser-aperture area and thus of the numerical aperture of the condenser. In this way there is achieved a convenient change between small luminous fields and large aperture areas on one hand, and on the other, large luminous fields and small aperture areas. Thus, with the whole set of objectives, there is immediate adaptation of the illumination conditions to what is required.

6.2 Reflectors for Reflected-Light Work

The function of reflectors has already been mentioned in Chapter 2, Section 2.1 as an essential part of the illuminator. The following four sections deal with the differences in the technical design of the reflectors.

6.2.1 Glass-Plate Reflector (Fig. 6.1)

The glass must be free from tension so that it does not introduce any polarising effect. Its front surface is coated with a film that increases reflection downwards towards the stage. The back of the reflector is coated with an anti-reflecting film so as to minimise the internal reflection downwards. When the beam returns

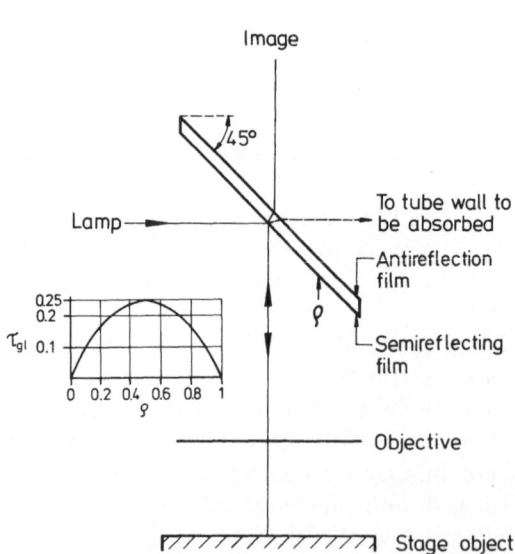

Fig. 6.1. Path of the axial ray in the glass-plate reflector.
Characteristic parameters: τ_{gl} transmittance of the glass-plate reflector (fraction) expressed as the ratio,

$$\frac{\text{Intensity of light leaving the glass plate in the direction of the image}}{\text{Intensity of light arriving from the lamp}}.$$

Neglecting the effects of reflection at the back face of the glass plate, absorption in the glass, and the interaction of the beam with matter outside the glass plate the transmittance is determined by the reflectance of the front face according to

$$\tau_{gl} = \rho(1 - \rho). \tag{6.1}$$

ρ reflectance of the front face (fraction) for the actual angle of incidence on the glass plate. In practice ρ is always smaller than 0.5. The inserted diagram shows τ_{gl} as a function of ρ

upwards from the stage object, there is considerable loss of light by reflection back towards the lamp and some absorption in the front-surface film; nevertheless, the overall eficiency of the coated glass plate is better than that of plain glass. According to Eq. (6.1) the efficiency is expressed by the total transmittance, and this has a maximum of 0.25, when the reflection-enhancing film has a reflectance of 0.5 and does not absorb. In practice the reflectance lies between about 0.3 and 0.7, with the result that the transmittance can only be up to about 0.2.

Natural light can be considered as a mixture of linearly polarised light having waves of equal amplitude and an infinte number of vibration directions. The waves having vibration direction parallel with the plane of incidence of the glass plate are called p-component, while those having the vibration directions normal to the plane of incidence form the s-component (German: senkrecht). The reflectance for the s-component is always greater than that for the p-component according to Fresnel's formulae for reflection at oblique incidence. These formulae are

$$\rho_s = \left(\frac{\sin(\alpha - \beta)}{\sin(\alpha + \beta)} \right)^2 ; \qquad (6.2) \qquad \qquad \rho_p = \left(\frac{\tan(\alpha - \beta)}{\tan(\alpha + \beta)} \right)^2 \qquad (6.3)$$

ρ_s reflectance for the s-component (fraction)
ρ_p reflectance for the p-component (fraction)
α angle of incidence
β angle of refraction. This is expressed with the help of Snell's law by $N = \sin \alpha / \sin \beta$

Example: For a simple glass-plate reflector α is 45°; taking $N = 1.5$ and the glass plate as being not coated; $\rho_s = 0.095$, $\rho_p = 0.0084$; taking $N = 2$ the values are $\rho_s = 0.21$, $\rho_p = 0.042$.

This has the following consequences:

1. On illumination with natural light the glass plate has a certain polarising effect.

2. On illumination with linearly polarised light, the vibration direction of the polariser being set East-West, the transmittance of the reflector as defined by Eq. (6.1) is greater than that if the polariser is set North-South, because in the former case the s-component, in the latter the p-component is effective.

3. The vibration direction of waves not lying in either the N-S or E-W planes is rotated by the glass plate. The result is a dark band running vertically across the image at observation between crossed polars (Galopin and Henry, 1972). Only within that band is rotation negligible, and only the area of the specimen lying within that band should be used for reflectance measurements with polarised light. Of course, the band is not visible when the analyser is out, as is the case during the measurement of reflectance.

In the sector of the light train containing the glass-plate reflector the marginal rays should be parallel (Figs. 2.1g and h) so as to avoid lateral displacements of the image and the appearance of parasitic (double) images (Fig. 12.3).

If the marginal rays are not parallel in the sector containing the glass-plate reflector, the primary image will be displaced laterally from its true position when the glass-plate reflector is in. With the rotating microscope stage this displacement is disadvantageous because any change in the reflector would necessitate re-centring of the objective or the stage.

The glass-plate reflector in no way restricts the aperture, and it provides symmetrical illumination; it may also be used to provide oblique illumination of a restricted aperture (Fig. 2.1d).

Note: The angle of inclination of the reflector should be preadjusted by the manufacturer. It should not be changed by the operator because any change in inclination would disturb the centring of the image of the luminous-field and of the condenser-aperture diaphragm. In certain designs of apparatus a small decentring due to this can be tolerated, provided that this can be corrected by moving the diaphragm in its own plane.

6.2.2 Mirror Plus Glass-Plate Reflector (Fig. 6.2)

The dark band mentioned in the previous section can be avoided by using the design shown in Figure 6.2. In this type (Smith reflector; Smith, 1964) a mirror is used to reduce the angle of incidence on the glass plate to 22.5°; this greatly reduces the rotation effect on the vibration direction of light waves mentioned in point (3) in the previous section. The result is that between crossed polars almost the whole field of view of the microscope is dark, hence the state of polarisation of light is almost uniform. The mirror has a reflectance of at least 0.9, hence the sacrifice of total transmittance as compared with the ordinary 45° glass-plate reflector is less than 10%; this is well worth while for the advantage just described.

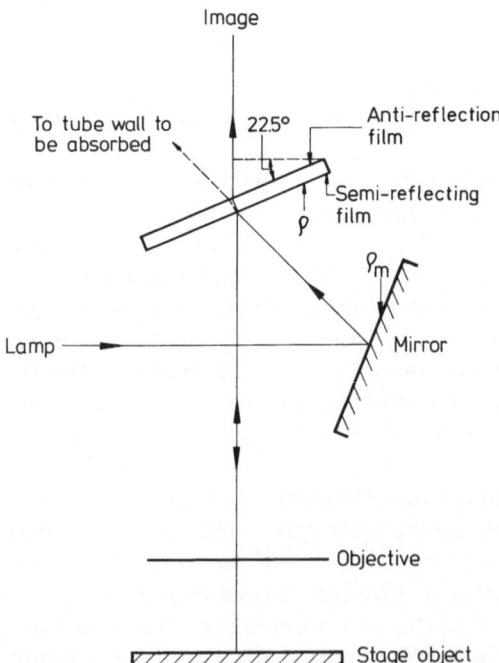

Fig. 6.2. Path of the axial ray in the mirror plus glass-plate reflector.
Characteristic parameters: $\tau_{m,pl}$ transmittance of the mirror plus glass-plate reflector (fraction) expressed as the ratio,

$$\frac{\text{Intensity of light leaving the glass plate in the direction of the image}}{\text{Intensity of light arriving from the lamp}}.$$

Neglecting the effects of reflection at the back face of the glass plate, absorption in the glass, and the interaction of the beam with matter outside the reflector the transmittance is determined by the reflectance of the mirror and of the front face of the glass plate according to

$$\tau_{m,gl} = \rho_m \cdot \rho(1 - \rho) \qquad (6.4)$$

ρ reflectance of the front face of the glass plate (fraction) for the actual angle of incidence on the glass plate; ρ_m reflectance of the mirror (fraction) for the actual angle of incidence on the mirror

6.2.3 Triple-Prism Reflector (Fig. 6.3)

The trapezium prism or Berek prism (Berek, 1936) achieves an almost entirely dark field between crossed polars, hence an almost uniform state of polarisation across the whole field for following reasons: In the prism the light undergoes three internal reflections. At each of these reflections there is a phase retard of the reflected beam compared with the phase of the incident beam; this retard is not the same for the two components, so that there is a resultant difference in phase of the p- and s-component. This difference increases with the refractive index of the glass. For index 1.74 it is about 60° [$(\pi/3)$ rad]; consequently the three reflections introduce a total phase difference of 180° (π rad), and a linearly polarised beam incident on the prism at any azimuth is rotated by 180° so that it is still linearly polarised on leaving the prism (for mathematical description see Rossi, 1957).

As the reflections are total no light is lost by reflection so that the prism has almost perfect transmittance [Eq. (6.5)]. In this it is greatly superior to

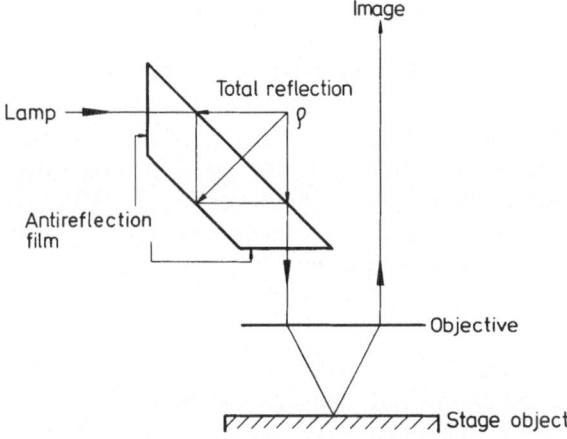

Fig. 6.3. Path of the axial ray in the triple-prism reflector.
Characteristic parameters: τ_{pr} transmittance of the prism reflector (fraction) expressed as the ratio,

$$\frac{\text{Intensity of light leaving the prism in the direction of the objective}}{\text{Intensity of light arriving from the lamp}}.$$

Neglecting the effects of reflection at the entrance face and at the exit face and of absorption in the glass the transmittance is determined by the reflectance of the effective prism faces as

$$\tau_{pr} = \rho^3 \tag{6.5}$$

(because of internal total reflection; $\rho = 1$)

Note: The glass has refractive index $N = 1.74$

the glass plate. The superiority can be expressed quantitatively as the ratio of transmittances [Eqs. (6.5) and (6.1)],

$$\frac{\tau_{pr}}{\tau_{gl}} = \frac{1}{\rho(1-\rho)} \tag{6.6}$$

τ transmittance (fraction)
pr   prism reflector
gl   glass-plate reflector
ρ reflectance of the front face of the glass plate (fraction). (Absorption in the glass and reflection at the rear face are neglected)

Taking the value of the denominator as 0.2 we see that the transmittance of the prism is greater than that of the glass plate by a factor of 5. In practice, the factor is mostly larger than this. But as compared with glass plates the prism has the disadvantage of filling only half of the aperture area of the objective acting as condenser; this is because the other half has to be used for the beam reflected upwards from the stage object. The result of this is to reduce to one half the aperture area of the objective and thus also the potential optical flux, so that the advantage of high transmittance of the prism is partly neutralised. Since for reflectance measurements only a small aperture is used, the intensity obtained with the prism is distinctly greater than with the glass-plate reflector.

Another disadvantage of the prism is that the light beam, falling on the stage object is oblique, with the effect that the field shows unilateral shadowing. This shadowing is the more distinct the more the aperture area of the objective is distant from the edge of the prism. But this disadvantage is compensated by the reduction of glare due to reflections at glass surfaces (lenses, etc.) situated between the prism and the stage object (App. 10). The prism prevents light arising from these reflections from reaching the image.

6.2.4 Chromatic Reflector (Fig. 6.4)

For the study of fluorescence phenomena in reflected light we need a reflector that deflects towards the stage object the short-wave exciting light and then transmits upwards the longer-wave fluorescent light emitted from it. This is done by a chromatic beam splitter which acts as a cut-off filter in reflection for illumination and as a cut-on filter in transmission, for imaging (Chap. 5, Sect. 5.3); (Ploem, 1967). This kind of beam splitter is available in several types having different typical wavelengths. The simple coated glass plate is cheap and easy to adjust, but the flanks of its spectral range of transmission or reflection respectively are not as steep as is sometimes desired. The flanks can be made steeper by combining the simple glass-plate beam splitter with an exciting filter of adequate pass-band shape placed in the illuminating train and with a barrier filter placed in the imaging train between the reflector and the image. Also a chromatic reflector of the mirror plus glass-plate type supplies steeper flanks.

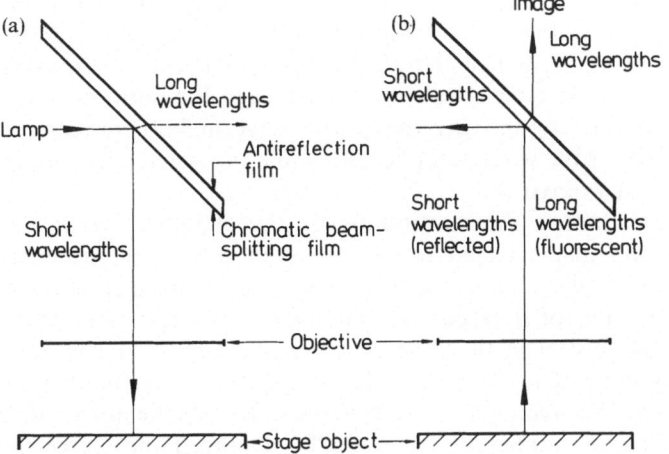

Fig. 6.4a and b. Glass-plate reflector acting as a chromatic beam splitter.

(a) Effect on the illuminating beam. The effective film on the glass plate deflects towards the stage object from the incident beam light having wavelengths shorter than a given threshold value and transmits light having wavelengths longer than the threshold value towards the tube wall where it is absorbed. The deflected light excites fluorescence. (b) Effect on the imaging beam. The effective film transmits fluorescent light having wavelengths longer than a given threshold value towards the image and deflects light having wavelengths shorter than the threshold value towards the lamp, where it is absorbed

6.3 Types of Microscope Stage

Microscope stages are divided into two main types; circular when rotation is required, as for use in polarised light, and square when it is not. The circular stage has to be mounted on precision ball or roller bearings because of the importance of accurate centring; this can be done by centring either the stage or else the objective. The stage may have a mechanical translation built in or else this may be added as an auxiliar stage. In all types of stages mechanical stability is essential. Large research microscopes allow interchanging of stages so that a choice may be made of the most suitable type for any particular purpose.

We consider the various types of auxiliary stage, and these can be used either in transmitted or in reflected light.

1. Temperature changing stage. This is usually referred to as a 'heating' stage, but the term is not very appropriate since it can also be used for cooling. The module is designed also for use with vacuum or with special atmospheres and at different pressures.

2. Mechanical stage. This is a simple type of hand-controlled stage having two mutually perpendicular motions and scales for setting the position. The mass to be moved is small, and so the setting can be done quickly and accurately. Of course, unless the mechanism is of high quality and the stage surface is flat, the specimen will go out of focus when moved.

3. Motor-controlled scanning stages. In the commonest kind the specimen holder itself is moved in one direction, while the whole stage carrying the holder is moved in the other. This has the disadvantage that the two masses are different, which can cause trouble when motors are used for both motions.

This difficulty is overcome by moving the whole stage in both directions, but the mechanism for this is somewhat complicated, while the speed of scanning is low on account of the mass involved.

The motors can be of either the continuous or the stepping types; the former provides continuous displacement while the latter moves the specimen in steps. Continuous motion allows a greater distance to be scanned in a given time and permits the making of a larger number of measuring values from different spots in the stage object in the time; but the more rapid the motion the less is the sampling time (Chap. 11, Sect. 8), i.e. the less the number of measuring values that can be averaged and then indicated in the form of a mean value. With decreasing number of values to be averaged the statistical spread of the mean for a given electrical noise from photometer modules is increased hence the precision of measurements diminished.

If we take the measuring area in the stage object to be as small as $0.5\,\mu m$ in diameter and this is to be scanned along a line of 1 mm in length, we would obtain 2000 measuring values; this movement could be done in some tenths of a second with a continuous motor, but in not less than ten seconds with a stepping motor.

4. Additional-axes stages. For use with circular stage on the microscope and polarised light in the study of crystalline specimens, the auxiliary stage is provided with one or more axes about which the specimen can be rotated. The kind with a single additional axis is called a spindle stage, while the others are called universal stages; the latter can have two, three of four additional axes.

6.4 Objectives

6.4.1 Properties

Objectives are characterised by the following properties:
   1. Magnification number (initial magnification, magnifying power) (MN_{Ob}); this lies in the range of 1 to 125.
   2. Numerical aperture (NA_{Obj}); this lies in the range of 0.04 to 1.3.
   3. Nature of corrections.

Note: In former times, the focal length was used instead of the magnification number. The relation between these is

$$f_{Obj} = f_T/MN_{Obj} \tag{6.7}$$

f_{Obj} the focal length of the objective (in mm)
MN_{Obj} magnification number of the objective (numeral)
f_T the focal length of the tube lens (in mm). This can be used with an objective that produces the image at infinity, and standard lengths are $f_T = 160$, 180, 200 or 250 mm. Eq. (6.7) can be formally applied also for an objective that produces the image at a finite distance. In this case

f_T is the corresponding tube length; this is normally 160 mm, and the relation is approximate only

We can set out the most important relations, and it should be noted that when the term 'numerical aperture' is used unqualified, the aperture on the stage-object side is intended.

1. The numerical aperture determines the limit of resolution, the depth of focus and the range of useful magnification [Eqs. (2.1) to (2.3)]. Objectives of the same numerical aperture, irrespectively of their degree of correction, have, theoretically, the same resolving power. Thus the improvement produced in the image by an objective of superior correction is due to better imaging and better contrast and not to better resolution.

2. For a given type of correction, the increase of the numerical aperture of an objective is less than linear with the increase of magnification.

3. For different types of correction the higher the correction the greater the potential numerical aperture at a given magnification.

Conditions (2) and (3) have the following consequences:

4. For an objective of given type of correction the numerical aperture on the image side (NA_{Obj}/MN_{Obj}) decreases with increasing magnification number or for an objective of given magnification number the numerical aperture on the image side increases with increasing degree of correction. Since, for a given field-of-view number of the ocular or a given diameter of the photometer diaphragm the potential optical flux of the objective is proportional to the square of numerical aperture on the image side, low-power objectives and those of higher degree of correction are able to supply a greater optical flux than high-power ones and those of lower degree of correction. The difference in optical flux due to different magnification number can be up to a factor of 6.5, that due to different degrees of correction up to a factor of 2.5.

5. The upper limit of the numerical aperture for dry objectives is 0.95, which corresponds to an angle of 70° as the maximum angle of incidence on the stage object ($NA_{Obj} = 0.95 = \sin 70°$). For an immersion objective with oil of refractive index of 1.5 the upper limit is 1.3 (in special cases 1.4) which also gives a similar maximum angle of incidence ($NA_{Obj} = 1.4 \approx 1.5 \cdot \sin 70°$) but has approximatively twice the optical flux for the same magnification number.

6. By using an immersion objective the numerical aperture is increased for a given aperture angle or for a given numerical aperture the angle of incidence is decreased; in addition there is enhancement of image contrast.

7. The higher the degree of correction, the more defective is the image when the objective is incorrectly used.

6.4.2 Corrections

We can set out the main types of aberrations for which objectives are corrected:

i. Monochromatic. This aberration is due to the separation of rays of the same wavelength passing through different points in the aperture area of the lens.

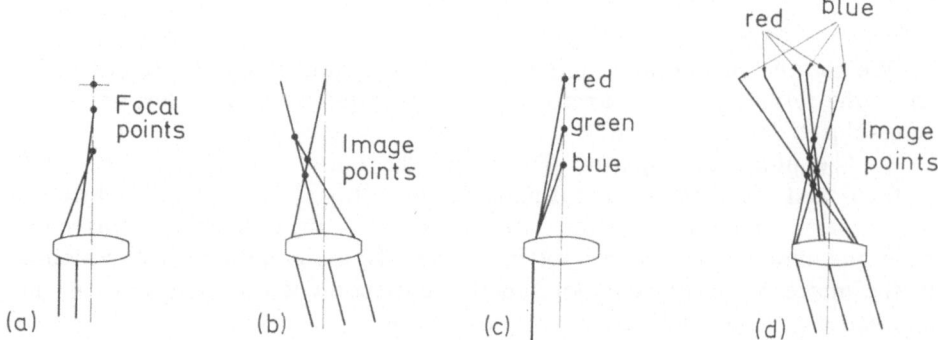

Fig. 6.5a–d. Basic aberrations of lenses.

(a) Spherical aberration (along the axis), aperture aberration. In a pencil of monochromatic parallel rays arriving parallel to the optic axis only those rays lying very close to the axis are refracted so as to intersect it in the focal plane. Rays lying further away from the axis are refracted so as to intersect it in planes closer to the lens. *Result:* the image of a point on the axis appears blurred.

(b) Asymmetrical aberration (normal to the axis). In a pencil of monochromatic parallel rays arriving inclined to the optical axis rays passing through different off-axial points in the lens are defracted so as to intersect mutually at different levels and different distances from the axis. *Result:* the image of an off-axial point appears as a comet-shaped blur.

(c) Chromatic aberration (along the axis). In a pencil of polychromatic parallel rays arriving parallel to the optical axis the monochromatic components of each ray separate out and so intersect the axis at different points. *Result:* the image of a white point in the axis appears with different coloration for different focusing positions of the lens.

Note: In a negative lens the effect is reverse.

(d) Chromatic aberration (normal to the axis). In a pencil of polychromatic parallel rays arriving inclined to the optical axis the monochromatic components of each ray separate out; rays passing through different points in the lens, as well as rays for different colours, form a network of rays with an infinite number of points of mutual intersection. These points are lying within a certain space. *Result:* the image of an off-axial point appears blurred and shows different coloration for different focusing positions of the lens

2. Chromatic. This aberration is due to the separation of rays of different wavelength on passing through the lens.

Both monochromatic and chromatic aberrations can be subdivided into a group referred to an image on the optical axis and a group referred to an image lying off the axis (Fig. 6.5).

3. Field curvature. This is due to off-axial points in the stage-object plane being imaged at levels below that at which the on-axial point is imaged, hence the focusing position of the objective must be changed if it is desired to have the off-axial point in sharp focus. (For general information about image defects see Meyer-Arendt, 1972.)

In practice several sources are, at the same time, involved in aberrations with the result that 'higher order' aberrations occur and it is extremely difficult to achieve perfect correction for all aberrations.

4. For numerical apertures greater than about 0.3 there is a distinct defect in image quality, due to aberrations in the cover glass (Fig. 6.6) that is used

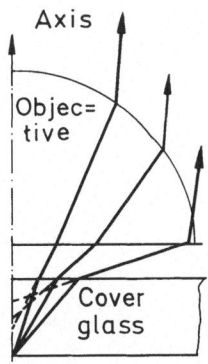

Fig. 6.6. Cover-glass aberration. Rays emanating from a point below the cover glass in the direction of the objective with different aperture angles are refracted at the upper face of the glass so that they appear to start from points at different levels within the glass along the axis. *Result:* the image of a point below the cover glass appears blurred

in standard transmitted-light microscopy. These must be taken into account in the correction of the objective. Actually the refracting power of the whole medium in which the objective is to operate is taken into account.

There are four main kinds of correction, and a properly corrected objective should always be used. Cover glasses are generally used in transmitted light work, while in reflected-light work the specimen is uncovered.

1. Dry, without cover-glass correction.

2. Dry, with cover-glass correction.

3. Oil-immersion without cover-glass correction.

4. Oil-immersion with cover-glass correction.

There are also kinds of correction for other liquids, e.g. water, glycerine, etc.

Notes:

1. In earlier times the cover glass and immersion oil were assumed to have the same refractive index (homogeneous immersion). This demand was abandoned by most manufacturers for several practical and theoretical reasons.

2. Standard cover glasses have a thickness of 0.17 with a tolerance of plus 0.01 or minus 0.03 mm and a refractive index of 1.542 ± 0.003 for the wavelength 546 nm. For special work (with heating stage, multiple-beam interference, etc.) the cover glass must be thicker, and special correction of the objective is required. (For the specification of standard cover glass and immersion oil, see German Standard DIN 58884.)

3. Just recently, special medium-power immersion objectives of the flat-field fluorite type were brought on the market. These can be used for water-, glycerine-, and oil-immersion in transmitted and reflected light in which the variation in correction with the change of immersion liquid is compensated by displacing certain lenses in the objectives along the axial direction by means of a control ring.

4. For a more detailed explanation of aberrations see Flügge (1956); Michel (1964) and Claussen (1967).

6.4.3 Main Types

We will now list the objectives in order of increasing quality in correction.

1. Achromats. Objectives of the simplest degree of correction are called achromats. In these, for two wavelengths the primary image is formed in exactly the same level; one wavelength must be less than 500 nm and the other larger

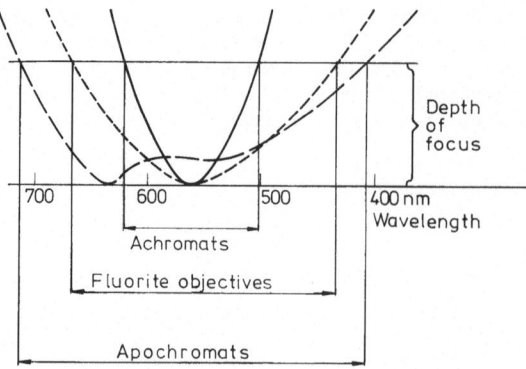

Fig. 6.7. Variation in focus of the objective on the stage-object side with variation in wavelength, schematic (after Brockhaus, 1961). We focus the objective onto a pointlike object lying on the optical axis so that a sharp image of this point is formed exactly in the nominal plane of the primary image with monochromatic light of the wavelength for which the monochromatic aberrations are corrected (wavelength at the point where a curve hits the abscissa). If the wavelength is changed and the level of the sharp image is to be maintained the object must be brought closer to the objective. In practice, the wavelength can be changed so far as the level of the object remains within the depth of focus so that no disturbance in the image is seen and the focusing position of the objective need not be changed. But when the wavelength range corresponding to the focal depth is exceeded the sharpness of the image is noticeably disturbed. The figure clearly shows that this wavelength range is the broader the higher the degree of correction of the objective

than 600 nm. For other spectral regions in the visible, the defective chromatic correction can cause noticeable displacement of the image along the axis (Fig. 6.7); this makes it necessary to change the focus position when changing the wavelength of monochromatic light. Monochromatic errors are corrected for one wavelength in the middle of the spectrum, usually for 550 nm.

2. Fluorite objectives. This name was originally applied to objectives in which low refracting glass is replaced by fluorite (CaF_2). They have similar correction to those just described but the material gives smaller residual chromatic aberration. This permits the numerical aperture to be larger and the variation in level of the primary image with variation in wavelength to be smaller than in an achromat. Further, fluorite objectives produce images with higher contrast than do achromats.

3. Apochromats. In apochromats the level of the primary image for a given level of the stage object or the level of the stage object for a given level of the primary image is made the same for three wavelengths (450 nm, 550 nm, and 650 nm approximately) with the result that the difference in focusing position of the objective for the whole visible (or even a larger) spectral range does not exceed the focal depth (Fig. 6.7); thus no false coloration is seen in the image. The level of the primary image is considered as being 'achromatic.' Monochromatic aberrations are corrected as for achromatic or fluorite objectives. Apochromats have an even larger numerical aperture than fluorite objectives of the same magnification number. Apochromats contain lenses of fluorite and other crystalline material.

All three kinds of objectives described above give a primary image in a curved plane. With additional corrections they can be made to give a flat primary image and this is indicated by the prefix 'flat-field' or 'plan.' Consequently the list of types of objectives must be supplemented by the following types:

4. Flat-field achromats.

5. Flat-field fluorite objectives.

6. Flat-field apochromats.

Flat-field objectives are particularly useful for large fields of view (ocular field) in which case they are combined with wide-angle oculars, for photomicrography and for all kinds of work in reflected light. They are not needed for microscope photometry because then only a small spot in the centre of the stage object or its image is measured.

6.4.4 Special Designs

Certain special designs are in widespread use in some fields, and we group them here merely because they have specific uses.

1. Polarised-light objectives. For use with polarised light the objective must be free from strain because otherwise it would depolarise the light passing through it. The materials are chosen to be of the highest purity and are worked, glued and mounted in such a way as to avoid any strain.

2. Phase-contrast objectives. These are equipped with a phase ring in, or near to, the back focal plane; this is made of a thin absorbing layer deposited from vapour on the lens, or else on a glass plate that can be exchanged. A ring-shaped aperture diaphragm in the condenser of appropriate size illuminates only the part of the aperture area of the objective that is occupied by the phase ring. The light waves traversing the phase ring are retarded in respect to those traversing the rest of the aperture area, and their amplitudes are weakened. The ring diaphragm causes illumination of the phase ring by undiffracted beams, and allows illumination of the rest of the aperture area only by beams that are diffracted at the stage object. Due to the proper phase and amplitude relations of diffracted and undiffracted waves and interference of these in the image plane, the primary image is seen with better contrast than it is with normal bright-field illumination.

3. Double-beam interference objectives. In these there is a birefringent crystal plate, either in front of the objective or else near its back focal plane. The crystal plate splits the light beam emerging from the stage object into two beams and these form two primary images in the same level but with a certain lateral distance from each other. The separation (or shearing) may be total, so that features with a certain width may appear twice and without overlapping or the separation may be partial, so that the distance between the split images of a feature is extremely small (differential interference). The contrast in the image results from the interference of the split light. With total shearing changes in optical thickness across the whole feature are made visible and measurable (Chap. 10, Sect. 2). For work in transmitted light the objective is combined

with a condenser that also has a birefringent plate. This is done in order to be able to use large numerical apertures. Without such a plate in the condenser the effective numerical aperture is restricted to values smaller than 0.1. With differential interference changes in optical thickness at the edges of features are made visible with better contrast than there would be with normal bright-field illumination owing to a shadowing effect which produces an illusion of three-dimensionality (Fig. 11.3 c).

Note: Of course, interference occurs only with a polariser and analyser crossed or parallel.

4. UV objectives. In order to allow the transmission of UV light (Fig. 6.8), these objectives are made either of a suitable crystalline material or else of special glass. They are chromatically corrected for the whole range between 250 and 700 nm so that they can also be used in visible light at the same focusing position. Those made under the trade name 'Ultrafluar' are available either as dry objectives or for immersion in glycerine, ordinary immersion oil being too absorbing in the UV range below about 310 nm.

5. Mirror objectives. These can be used for UV light as well as for visible and IR light, but they are not really suitable for use in reflected light because the central part of the light cone is lacking (Fig. 6.9). Neither of these are suitable for the quantitative study of the polarisation behaviour of the stage object because the reflections cause depolarisation; in addition only oblique rays are effective for imaging.

A combination of mirrors and lenses may be used. There is no need for particular chromatic correction, but they must be used with cover glasses of the correct thickness. They do produce monochromatic errors and it is extremely difficult to correct for these. They have a large working distance and so are convenient for the manipulation of the stage object during observation or measurement.

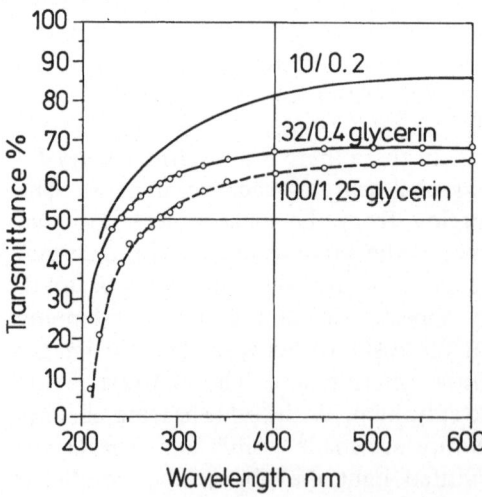

Fig. 6.8. Spectral-transmittance curves for UV objectives. Type of objective: Ultrafluar

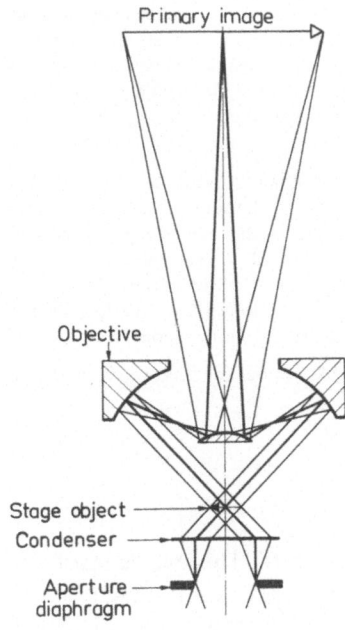

Primary image

Objective

Stage object

Condenser

Aperture
diaphragm

Fig. 6.9. Path of rays in a mirror objective, schematic. The objective is taken as consisting of a concave ring mirror and a convex mirror in its centre having the shape of a sphere cap. The figure shows the path of rays starting at the condenser-aperture area and ending at the plane of the primary image. The rays accepted by the objective form a ring cone with parallel inner and outer faces and the top being the stage-object field. It is clear that the stage object is traversed only by oblique rays. The thick straight lines are rays that intersect the optical axis in the centres of the stage object and primary image, the thin straight lines are rays that intersect at two opposite points on the edge of the fields. The points at which all the rays intersect are lying on the central circle in a ring, which forms the effective aperture areas of the condenser and the objective

Notes:
1. Objectives that are to be exchanged must be designed for the same standard lengths in the microscope; in this the total distance from the stage object to the primary image and certain sectors in this distance are involved, not only tube lengths (Fig. 6.12).

2. With glass or crystal lenses an achromatic level (Fig. 6.7) is achieved only by tolerating a certain secondary effect, which is the chromatic difference in magnification (Fig. 6.10). This effect is superposed on that of chromatic aberrations along, or normal to, the axis (Fig. 6.5a and b) and it can be seen quite distinctly. It is neutralised by combining the objective with a compensating

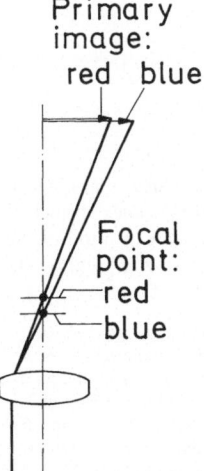

Primary
image:
red blue

Focal
point:
red
blue

Fig. 6.10. Chromatic difference in magnification. This kind of aberration is due to different focal lengths and different distances of the focal planes from the lens (objective or ocular) for different wavelengths. The image appears sharp at the same focus but is of different size for different wavelengths; the magnification number of the objective may be up to 2% larger for blue light than for red light. Hence a blue rim is seen at the edge of a white feature in the primary image on the side of the ocular diaphragm and a red rim on the opposite side, the width of the rims being proportional to the distance of the rim from the centre of the field in the ocular. Of course, the coloration is seen in the ocular, unless the latter has a compensating effect.

Note: Chromatic difference in magnification must not be confused with the basic chromatic aberrations (Fig. 6.5c and d) ocular or compensating lens that reverses the same amount of chromatic difference in magnification. Unfortunately, up till now, there is no agreement between different manufacturers about a standard amount of chromatic difference in magnification, so that perfect compensation is achieved only when all the optics are made by the same manufacturer. As the effect is negligible in the centre of the field, any observation or measurement limited to this is unaffected; also, operation with monochromatic light is unaffected.

3. Although all objectives previously described are corrected for imaging with visible light (UV objectives also for imaging with UV light) they can with a certain restriction also be used for imaging with near IR light (up to about 1100 nm). The restriction is the use of a not too large spectral bandwidth, hence always to operate with a monochromatic filter that selects a spectral band of IR light. When the wavelength is changed, the focusing must be corrected. Of course, also in the IR range, objectives of higher degree of correction are superior to those of low degree. For still larger wavelengths special design of all optics, hence a special 'microscope' is required.

4. The objectives are transmittant for wavelengths up to about 2000 nm but for such wavelengths the correction is insufficient.

6.5 Oculars

The ocular is made of two lenses (single or compound). The eye lens of the ocular magnifies the primary image in vision, hence acts as a magnifier of which the primary image is the object and converts this into a virtual image. For photography or projection it is adjusted so that a second real image is formed (Chap. 2, Sect. 3.1). The second real image may also be formed in the plane of a separate photometer diaphragm. The field lens of the ocular collects the light coming up the tube and together with the eye lens of the ocular forms an image of the aperture area of the objective in a plane where the iris of the observer or a photocathode that has a small diameter can be conveniently placed (exit pupil of the microscope).

There are two main types of oculars (Fig. 6.11). In the Huygens type the primary image is formed between the two lenses and it is here that must be

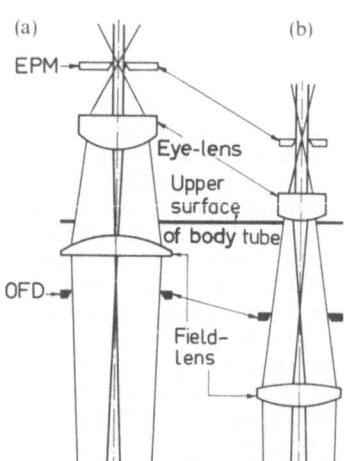

Fig. 6.11a and b. Lenses and ray paths in oculars. *Thick straight lines:* marginal rays; *thin straight lines:* principal rays. OFD Ocular diaphragm; EPM exit pupil of the microscope. (a) Ramsden ocular (front diaphragm); (b) Huygens ocular (internal diaphragm).

Note: (b) shows the effect of the field lens on the primary image when this lens is below the nominal primary-image plane; the primary image is displaced to a lower level than that in the Ramsden ocular, simultaneously the diameter of the image is diminished. Hence, the ocular diaphragm is in a lower level and has smaller diameter, while the length in the stage object corresponding to the diameter of the diaphragm is the same for both oculars placed the ocular diaphragm and the cross wires, grating or scales; this type is mostly used for oculars of low magnifying power. In the Ramsden type the primary image is formed below the field lens and so the cross-wires, etc., are placed there; this type is mostly used for oculars of higher magnifying power. Oculars may be designed with compound lenses, so as to compensate the chromatic difference in magnification, due to the correction of the objective (Note 2 of Chap. 6, Sect. 4.4). Because this difference mainly exists for objectives with high degree of correction, compensating oculars are mostly used with high quality objectives. The observer can easily check whether the ocular has a compensating effect or not; in compensating oculars the field diaphragm shows a yellow, in non-compensating ones a blue seam.

The ocular is characterised by two numbers:

1. The magnification number (MN_{Ocl}) is the angular magnification under which the primary image is seen as a virtual image with the ocular; alternatively it is the factor by which a length in the second real image is greater than the corresponding length in the primary image, if the ocular acts as a projecting lens and the second image is formed at a distance of 250 mm from the exit pupil of the microscope. Standard magnification numbers are between 5 and 25.

2. The field-of-view number (FN) is given in mm. In a Ramsden ocular (Fig. 6.11 b) in which the ocular diaphragm is situated in front of the whole ocular-lens system the field-of-view number is the same as the diameter of the diaphragm. In a Huygens-type (Fig. 6.11 a) in which the ocular diaphragm is situated between the field lens and the eye lens of the ocular the field-of-view number is greater than the diameter of the diaphragm. The reason for this is that the field lens has the effect of displacing the primary image into a level below that in which the image would be formed without the effect of the field lens. The displacement results in diminishing the magnification scale of the primary image relative to the scale obtained without the effect of the field lens. In order to maintain the diameter of the field in the stage object selected by an ocular with front diaphragm, i.e. in order to maintain the field-of-view number, if the ocular with the front diaphragm is replaced by an ocular with internal diaphragm, the diameter of the internal diaphragm must be smaller than that of the front diaphragm. If, on the other hand, the diameter of the diaphragm in the Huygens ocular is the same as that of the diaphragm in the Ramsden ocular the Huygens ocular has a larger field-of-view number than the Ramsden ocular.

Oculars to be inserted in body tubes of standard inner diameter (23 mm) have field-of-view numbers between 6.5 mm (high-power oculars) and 20 mm (low-power oculars). For a given magnification number oculars to be inserted in wide body tubes have field-of-view numbers up to 25% larger than those with standard diameter.

Notes:
1. The diminishing of the primary image by the effect of the field lens of the ocular is, of course, taken into account for the magnifying effect of the eye lens, so that for a given magnification number the eye lens of a Huygens ocular has a stronger magnifying effect than the eye lens of a Ramsden ocular.

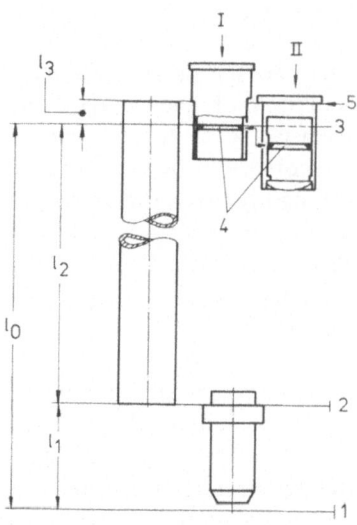

Fig. 6.12. Standard lengths in microscopes. These lengths are agreed as standard by most microscope manufacturers and are given in the German Standard DIN 58887.

Definitions and values of lengths: l_0 Distance between the object and the primary image = 195 mm or ∞; l_1 Parfocal distance of the objective (Distance between the object and the contact surface of the objective) = 45 mm; l_2 Primary-image distance of the objective = 150 mm or ∞; l_3 Parfocal distance of the ocular = 10 mm.

Planes: *1* Stage-object plane. *2* Contact surface of the objective. (Lower surface of the nose piece.) *3* Primary-image plane of the objective, being in the same level as the *4* Primary-image plane of the ocular. *5* Contact surface of the ocular. (Upper surface of the body tube.) *I* Ocular with front diaphragm. *II* Ocular with internal diaphragm.

Notes:

1. The sum of the distances l_3 plus l_2 is the same as the tube length (160 mm or ∞). But this term is no more used as a standard term.

2. The values of the lengths are nominal. The actual lengths, for example the distance between the object and the primary image can have other values: but differences between nominal and actual values must be optically compensated.

3. Objectives with correction for infinite object-image distance must be combined with a positive tube lens. This lens projects the primary image in the plane having the nominal primary-image distance in the ocular.

4. In the British Standard 3836 a tube length of 160 mm but a primary-image distance of the ocular of 15 mm is recommended. Although the tube length is the same in the German and British Standard the combination of optics designed according to either recommendation is not possible without defective adjustment and image quality

2. The manufacturer assumes a certain distance of the primary-image plane from the uppermost surface of the body tube, and for this 10 mm is taken as agreed by most manufacturers (Fig. 6.12); only oculars with the same primary-image distance should be interchanged on a microscope. (See also British Standard 3527.)

From the magnification number (MN_{Ocl}) and the field-of-view number (FN) we can derive the following important parameters,

1. Diameter of field in the stage object $= FN/m$ (6.8)

m = magnification scale of the primary image, resulting from the product of magnification number of the objective and the tube factor

2. Focal length of the ocular $= 250$ mm$/MN_{Ocl}$ (6.9)

3. Field-of-view angle (w) of the ocular; $\tan\left(\dfrac{w}{2}\right) = \dfrac{FN \cdot MN_{Ocl}}{2 \cdot 250 \text{ mm}}$.   (6.10)

High Eye Point Oculars. In ordinary oculars the distance from the uppermost surface of the eye lens to the exit pupil of the microscope is about 5 mm. In high eye point oculars this distance is 15 mm; this is very convenient for observers with spectacles while a photocathode may also be situated in this plane.

Comparison-Pattern Oculars. All oculars having micrometer scales or grids, patterns of any kind, or even an iris diaphragm come under this category; all oculars for use with polarised light and circular stages possess at least cross-wires. In all these oculars the eye lens is focusable.

Projecting Lenses. For photomicrography, projection on a screen, television of microscopic images or for work with a photometer head with the photometer diaphragm at a level distant from that of the primary image it may be necessary to produce a second real image in a suited level. This is achieved in one of the following ways: (1) by displacing the ocular as described in Chapter 2, Section 3.1, (2) by placing a positive lens in or close to the exit pupil of the microscope, (3) by replacing the ocular by a negative lens and placing this lens below the plane of the primary image.

The ratio of a distance in the second real image to the corresponding distance in the primary image is called projection factor, or in photomicrography it is called camera factor.

6.6 Body Tubes and Tube Optics

The body tube is the metal housing having the objective at its lower end and the ocular at its upper side; it encloses the optical train between these two lenses. In modern designs it is made up of several modules, each of which can be interchanged with another of corresponding function. These modules are: objective holders (for a single objective, or a revolver for a set of objectives), intermediate tubes (optical-mechanical devices with slits for auxiliary optics, Bertrand lens and diaphragm, etc.) and tube heads. The auxiliary optics here comprise an analyser, compensators, beam splitters, and tube lenses. The heads comprise optics for viewing (monocular or binocular), for photometry, for photography, and for projection.

Since the mechanical requirements do not usually permit the actual tube length to be the same as the nominal length used in the calculation of lenses, the difference is compensated by tube lenses; these displace the primary image just by the length by which the actual body tube is longer than the nominal one. Tube lenses may produce a sector in the light train in which the marginal rays are parallel (Fig. 2.1 g); in this may be placed any parallel-sided optical element without disturbing either the level or the quality of the primary image. There is, however, a limit to this variation of the level of the actual primary image. For an increase of the actual tube length the exit pupil of the microscope is brought closer to the eye lens of the ocular, which produces the 'keyhole' effect; this can be brought about with high eye-point oculars if the actual tube length is too great.

Such displacements may or may not change the scale of the primary image. The tube factor is the ratio of the actual scale to that which would be produced without the tube lenses or the ratio of the focal lengths of the positive and the negative tube lens; it is usually engraved on the tube module.

A 'wide-field' tube lens can be used to increase the size of the stage-object field that can be viewed; such a lens has a 'magnification factor' less than unity, so that there is a loss of magnification of the primary image.

A tube lens can be made with a compensating effect, so that the primary image is already corrected for chromatic difference of magnification; in such cases it is not necessary to use a compensating ocular to obtain the best image in the eye.

A set of interchangeable tube lenses can serve for stepwise changes of magnification of the primary image. A pancratic tube-lens system (zoom system) can serve for continuous change in magnification, thus making unnecessary to change the oculars for this purpose. But this kind of change in magnification has no influence on the effective numerical aperture and thus none on the resolving power of the objective or on the useful range of magnification.

CHAPTER 7

The Photometer System

7.1 Photometer Heads

The arrangement of main modules in the microscope photometer is given in Figure 4.1 where it is seen that the photometer system in the literal sense of the word comprises everything between the microscope and the device that signals or records the light amplitudes in the form of measuring values. We begin with the photometer head, which is a mechanical interface connecting the body tube of the microscope to the photosensor.

A great variety of heads is now on the market, and it is easy for the buyer to become confused among the arguments in the sales literature. Here we can only describe the various factors involved, so that the buyer can at least clear up his mind on what he wants to get. What he can have depends on how much he is willing to pay for convenience in making the measurements, so let us list the essential facilities for doing this accurately, as far as the head is concerned.

1. A photometer (measuring) diaphragm in the plane of a real image for selection of the area to be measured on the stage object.

2. Optics for viewing the photometer diaphragm, either directly or as represented by previously-adjusted marks.

3. A support enabling the photocathode to be positioned close to the exit pupil of the microscope (or to a plane conjugate to this); alternatively it may be positioned close to the primary image (or to a plane conjugate to this), unless the diameter of the area to be measured is too large.

Note: The conditions that dictate the positioning of the photocathode are:

1. The light cone falling on the cathode must not exceed the area of this; if the cone is small, then it may be increased by means of a lens or a diffusing glass so as to spread the light over a larger area of the cathode. The area should not be less than about 1 mm in diameter.

2. On account of varying sensitivity over the cathode layer, the light spot should be kept in the same place during the measurement of specimen and reference material or during the whole series of measurements that is related to the same reference measuring value.

3. Where an image converter or a television camera tube is in use, their surfaces always have to be accurately in the plane of a real image.

The simplest type of photometer head is merely a housing placed on top of the ocular so that the photocathode lies exactly in the plane of the exit pupil of the microscope. This housing has to be removed while the area to be measured is selected by the operator looking down the vertical tube through the ocular and using the ocular diaphragm as photometer diaphragm or replacing the ocular diaphragm by a photometer diaphragm of smaller diameter or particular shape. This device is least expensive but it is rather inconvenient to operate.

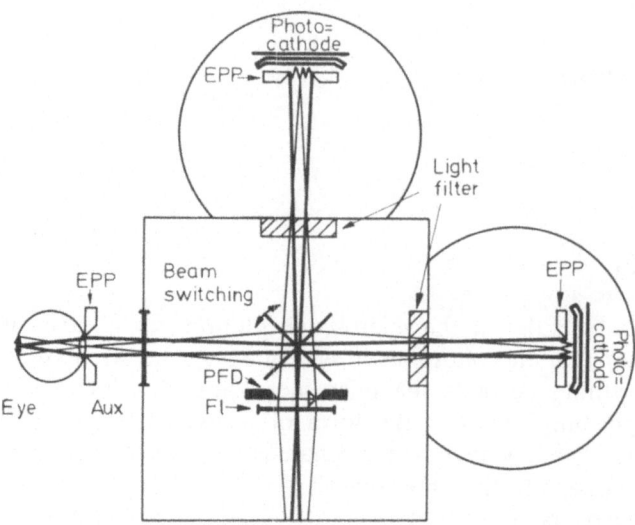

Fig. 7.1. Facilities in photometer heads of standard design. The figure shows a type of photometer head which allows two photosensors or one photosensor and any other device (such as a photo-camera) to be used alternately. The thick straight lines show the path of marginal rays, the thin ones that of principal rays.

Abbreviations: EPP Exit pupil of photometer head; Aux Auxiliary lens; FL Field lens; PFD Photometer (field) diaphragm.

The photometer diaphragm selects the portion to be measured of the real image of the stage object; the viewing lens enables this portion to be seen. The beam deflector or splitter permits the choice of ray paths: (1) for viewing the photometric diaphragm, or (2) for transmitting the light beam to one or other of different photosensors. The white-light image of the stage object can always be observed, no matter what filter may be inserted in front of the photocathode

At somewhat higher cost this housing can be placed on a split body-tube device having a straight tube for the photocathode and an oblique tube for viewing during selection of the area to be measured (photometric field). The ocular in the viewing tube may contain a concentric ring reticule, each ring being calibrated to represent the measured area of each of the photometer diaphragms. In more costly photometer heads additional conveniences and facilities exist, and these are set out in Figure 7.1. These include: (1) Exchangeable or variable photometer diaphragms and optics for viewing the diaphragms and the real image in the plane of the diaphragms, (2) the possibility of measuring and observing in different wavelength ranges, UV, visible and IR, and (3) the use of area- and wavelength-scanning devices.

The most convenient of all is the type of head that projects the photometer diaphragm, hence the photometric field, in the primary image plane (viewed through the ordinary ocular); in this way the same viewing system is used for everything.

Image-area scanning is achieved by means of a photometer diaphragm moving laterally over the image, or else by a mechanism placed in the plane of the objective-aperture area (or conjugate plane) that makes the marginal rays to intersect at different points in the image plane; thus to move the image across a fixed photometer diaphragm (Chap. 9, Sect. 2).

7.2 Photomultiplier Tubes

7.2.1 Side- and End-Window Types

Photomultiplier tubes of standard design have so far proved to be better suited for photometric work with the microscope than other kinds of photosensors. Hence, only photomultiplier tubes will be taken into consideration in this book. But it should be mentioned that the technical design of other kinds of photosensors, e.g. photodiodes, photoresistors or photoelectron counters (electron channel multipliers), is rapidly improving so that these also may, in future, be useful for microscope photometry. There are several structurally different types of photomultiplier tubes. The two types shown in Figure 7.2a and b are the ones mostly used.

The side-window type is quite small and easily attached to the photometer head. This type and its power supply unit are less costly than the end-window type, which can have more dynodes than the first type and hence can reach a greater sensitivity; also the end-window type, can have less dark current and electric noise. It is this type that can be used with fibre optics that conduct light from a part of the light train to the photosensor lying outside this train, because the fibres can be brought in close contact with the photocathode.

For further technical information on photomultiplier tubes the reader is referred to Greif (1972), Young (1972) and to manufacturers' catalogues. These contain all instructions needed for correct operation with these tubes.

7.2.2 Response Characteristics

Most photomultiplier tubes carry a number with the prefix S, and this is agreed internationally; the S-number defines the material of the photocathode and the spectral response of the tube. But there are some tubes that have no S-number and which are labelled only with the manufacturer's code. Improvements in the design of photomultiplier tubes are being made rapidly, leading to an increase of spectral sensitivity and range.

In practice it is the material of the window that sets a limit at the short-wave end. For ordinary glass windows this occurs at about 320 nm, but special materials transmit radiation down to about 250 nm. The range below this is not important for microscope photometry. The long-wave limit is set by the nature of the photocathode itself (Fig. 7.3 and Table 7.1); it can be seen in the figure that, for example, the S-5 cathode fades out between 600 and 700 nm, and this type should not be used for measurements where the whole of the visible range is of interest.

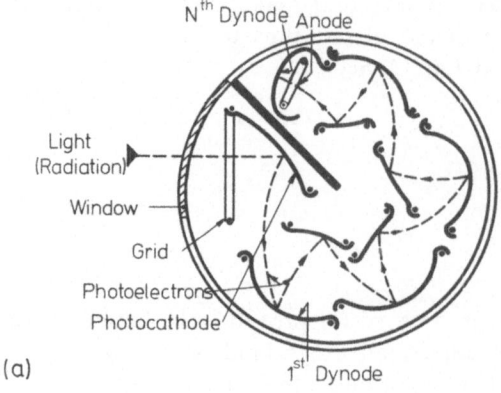

(a)

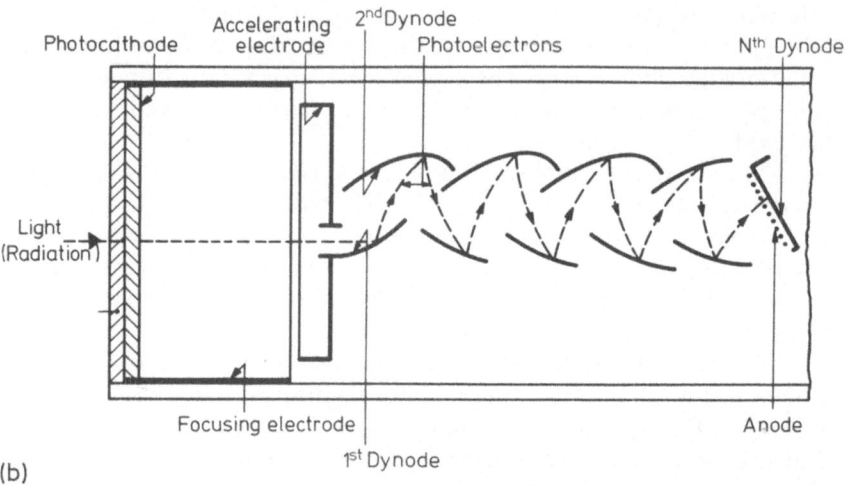

(b)

Fig. 7.2a and b. Technical design of photomultiplier tubes. A photomultiplier tube consists of an evacuated transparent envelope containing a photocathode, an electrostatic focusing system to constrain the electrons into paths with little spread between dynodes, a multiplier system (dynodes) and an anode. When light falls on the cathode, electrons are released; these are then multiplied by the dynodes, each of which is at a positive potential to the previous one. At the anode the current is used to activate the indicating device.

(a) Side-window tube. This tube has a compact focused structure. The light falls from the side on a reflecting cathode, the number of dynodes being usually limited to 9. (b) End-window tube. This tube has a linear focused structure. The light falls from the front on a semi-transparent cathode, the number of dynodes being up to 14.

Note: There are several modifications in the arrangement of dynodes in this type, e.g. as in the 'venetian-blind type'

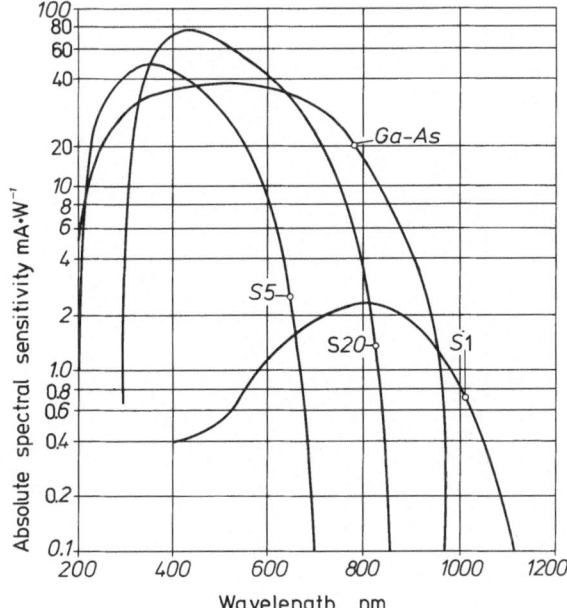

Fig. 7.3. Typical absolute spectral sensitivity of photocathodes. *Curves:* types of photocathodes that can be recommended for use in a microscope photometer. The cathode material is specified in Table 7.1.

Note: There are several variants of these types. In most variants the useful spectral range of sensitivity is extended towards longer wavelengths with the consequence that the absolute maximum sensitivity is decreased but the integral sensitivity remains the same

A new type of photocathode is being developed made of GaAs and having a fairly uniform spectral response over a very wide range (Fig. 7.3). This is a useful property for measurements with data processing apparatus (wavelength scanning), but, at the present time the whole photomultiplier system linked to this cathode has an overall sensitivity that is much less than that of a conventional tube. Improvement in sensitivity in future can be expected.

Statistical variation is experienced in production, and each photomultiplier tube should be individually selected for use with the microscope, both for sensitivity and for linearity of response. The sensitivity varies over the area of the cathode by about 1%; in side-window tubes this variation is, in the average, greater than in end-window tubes. The sensitivity varies also with the direction of vibration of polarised light and if the vibration direction is turned round the axis of the light beam, a variation up to 30% can be expected for side-window tubes but only up to 5% for end-window tubes.

Note: In all cases the photocathode in a microscope-photometer system should be properly aligned in the light train so that it supplies the greatest possible photocurrent for a given radiant flux. For this purpose it should be moved across the photometric beam and rotated into all directions until, for a given setting of the microscope, the maximum measuring value appears. At this position it should be fixed and operated.

The *absolute sensitivity* (or response) of the photocathode is the photoelectric current obtained from the cathode by a radiant flux of unit amount; the absolute sensitivity varies with wavelength as shown in Figure 7.3.

The *gain* is the internal amplification in the tube achieved by the dynode system; it is the ratio of the overall sensitivity of the photomultiplier to the

Table 7.1. Technical data of selected cathodes in photomultiplier tubes

Type of cathode	Cathode material	Window material	Cathode radiant sensitivity (S_k)[a] [mA/W]	Wavelength of maximum response λ_{max} [nm]	Useful spectral range nm	Cathode dark current [A/cm²] at		Radiant-flux input equivalent to the cathode dark current [W/cm²][a] at	
						+20°C	−20°C	+20°C	−20°C
S 1	Ag, O, Cs	Borosilicate or lime glass	2–3	800	420–2280	10^{-12}	$5 \cdot 10^{-15}$	$3\text{–}5 \cdot 10^{-10}$	$1.7\text{–}2.5 \cdot 10^{-12}$
S 5	Cs, Sb	UV glass	50	350	220– 600	10^{-17}	10^{-19}	$2 \cdot 10^{-16}$	$5 \cdot 10^{-19}$
S 20	Na, K, Cs, Sb (Multi-alkali)	Borosilicate or lime glass	75	410	320– 790	10^{-16}	$5 \cdot 10^{-19}$	$1.5 \cdot 10^{-15}$	$7 \cdot 10^{-18}$
(Ga, As) (no S-number)	GaAs	UV glass	40	600 ± 200	200– 900	10^{-16}	No value available	10^{-15}	No value available

[a] The quantity is specified for the wavelength of maximum response.

sensitivity of the photocathode alone,

$$G \equiv S/S_k \tag{7.1}$$

G gain (numeral)
S overall sensitivity of the photomultiplier; anode current per unit of radiant flux (in A/W)
S_k photocathode sensitivity (in A/W)

Within the range of standard operating voltages the gain lies between 10^5 and 10^7.

There is a linear relation between relative change in supply voltage and gain, the proportionality factor being approximately the number of dynodes. This is expressed as

$$\Delta G/G \approx N \cdot \Delta V/V \tag{7.2}$$

ΔG change in gain (numeral)
ΔV change in supply voltage (in V)
G gain (numeral)
V supply voltage (in V)
N number of dynodes

Note: The change in gain is a function of the supply voltage as it is the gain; thus $\Delta G/G = f(V)$.

If we take N as being 10, then it is clear that the stabilisation of the supply voltage has to be ten times better than the constancy demanded of the anode current.

In some tubes the sensitivity diminishes (*after-effect*) after a short interruption of illumination, but the complex reasons for this kind of behaviour are not yet known. Another effect is a slow but continuous decrease of sensitivity (*fatigue*) while the tube is at high voltage, and this occurs with the cathode unilluminated; the effect is reversible. Irreversible effects are called '*ageing*,' and this can occur when the tube is not under voltage. In general, all that can be done is to select tubes free from the first of these effects and to prevent it being strongly illuminated in order to reduce the other two effects.

Temperature, of course, affects all electrical modules, and, during operation, a constant temperature should be maintained. Cooling is beneficial in reducing both dark current and noise; for this purpose thermoelectric coolers are available.

Too high supply voltages are harmful to photomultiplier tubes. For side-window tubes the voltage can be up to about 1200 V, for end-window tubes up to about 2500 V. Great stability of voltage is, of course, essential for the reason described above. The supply current is in the magnitude of mA. Recently power sources have been put on the market, that can be immediately linked to the tube socket, hence removing difficulties in the insulation of high-voltage circuits.

7.2.3 Dark Current

Dark current is a permanent electric direct current in the tube caused mainly by thermionic emission from the cathode and neighbouring electrodes; the amount of dark current depends on the thermal work function of the cathode, the size of this, and the temperature. Minor contributions to the dark current come from isolation currents, field emission at edges of electrodes, ionisation of residual gas, and also from radioactivity in the materials of the window, and cosmic rays.

The dark current can be expressed in the unit of A or, alternatively, in that of W, this being the unit of the radiant flux of light that would produce a photocurrent of the same amount as that dark current.

The cathode dark current is related to the area of the cathode; the anode dark current is approximately greater by the gain factor. The cathode dark current increases with the long-wave limit of the spectral response and also with the ambient temperature; consequently the S-1 type which covers the near IR, has the largest dark current, and cooling is especially beneficial with this type (Table 7.1).

Dark current electrons are emitted from the whole surface of the cathode, not only from the illuminated part; it follows that the total area of the cathode should not exceed by much the area on which the light beam is going to fall. There are designs of tubes that permit only electrons from a small central area on the cathode to enter the multiplying system; these are very suitable for photometry with the microscope where the cathode is placed at a level where the photometric beam is very narrow.

The dark current is largely unaffected by the supply voltage, except that a very high voltage ionises residual gas. When the cathode is exposed to too much light, energy is trapped in the cathode, and it takes some time for this to be dissipated by the emission of dark current electrons; the application of high voltage at this time causes very high dark current. Several days without exposure to light is required to restore the value of the dark current to its initial value. The application of high voltage while the cathode is exposed to strong illumination results in its destruction.

The dark current causes incremental measuring values, the effect of which is described in Appendix 5. But with a tube of high quality the contribution of dark current to the actual measuring value is generally small, unless the photocathode operates close to the limit of spectral response (see note in Chap. 12, Sect. 3.4). The dark current can be ignored if it is less than 1% of the photocurrent. If it is more than this, the following steps can be taken:

1. Cooling the tube.

2. Insertion of a special electrical compensating circuit.

3. Setting the indicating device to a negative reading (measuring value) at its zero point equivalent to the dark-current reading.

4. Substraction of the measuring value for dark current from each actual measuring value.

5. Operation with modulated light (Chap. 4, Sect. 5).

7.2.4 Noise

A distinction has to be made between dark current and noise. Noise comprises all the unwanted electrical impulses from equipment components, by man-made interference and by natural disturbances. In photometry the noise produces an uncertainty in the final measuring value. Among the sources of equipment noise are the following:

1. lamp flicker,
2. disturbance of the light train by vibrations of a mechanical nature,
3. fluctuation of electric supply,
4. statistical spread of electron behaviour in electrodes, conductors, and semiconductors,
5. photomultiplier noise.

It is from the fluctuation of photon input and of dark current and other electrical emissions in the tube that photomultiplier noise arises, and this can be expressed as the standard deviation of the mean output current in A or else in W of the equivalent input light (radiant flux). As the rule, the noise arising in the photomultiplier tube does not exceed 10% of the dark current but that arising in other modules is much greater.

Noise in the whole measuring equipment is proportional to the square root of its 'display-band width,' which is the range of frequency of changings in measuring values the indicating unit is able to display. It is expressed in units of $s^{-1} = Hz$. The reciprocal of this is the sampling time or time constant. (For definition of sampling time see Chap. 7, Sect. 3.1)

The less the noise, the greater is the precision of the measuring values. A ratio of at least 3 is required between the measuring value for the stage object and that for the noise, for the measuring value to be accepted as true, while the ratio rises to 10, before the measuring value can be considered reliable.

It is clear that at high scanning speeds the disturbance by noise is greater than at low; a compromise has to be reached between speed of scanning and reliability of the measuring value.

For the reduction of photomultiplier noise the same steps can be taken as were taken (Sect. 2.3) for the reduction of dark current; further, steps should be taken to ensure the highest stability—electrical and mechanical—of the whole measuring equipment.

Note: The electron-transit time in the photomultiplier tube is about $2 \cdot 10^{-8}$ s for a 10-dynode system and about $5 \cdot 10^{-8}$ s for a 15-dynode one. Another factor is the time of anode-pulse rise, which is in the order of 10^{-9} s.

7.3 Electrical Conversion Modules

7.3.1 Important Functional Properties

Electrical conversion modules comprise all modules that convert the radiant flux to a measuring value as shown in Figure 7.4. The radiant flux may be produced with modulated light or with direct light. The nature and the use

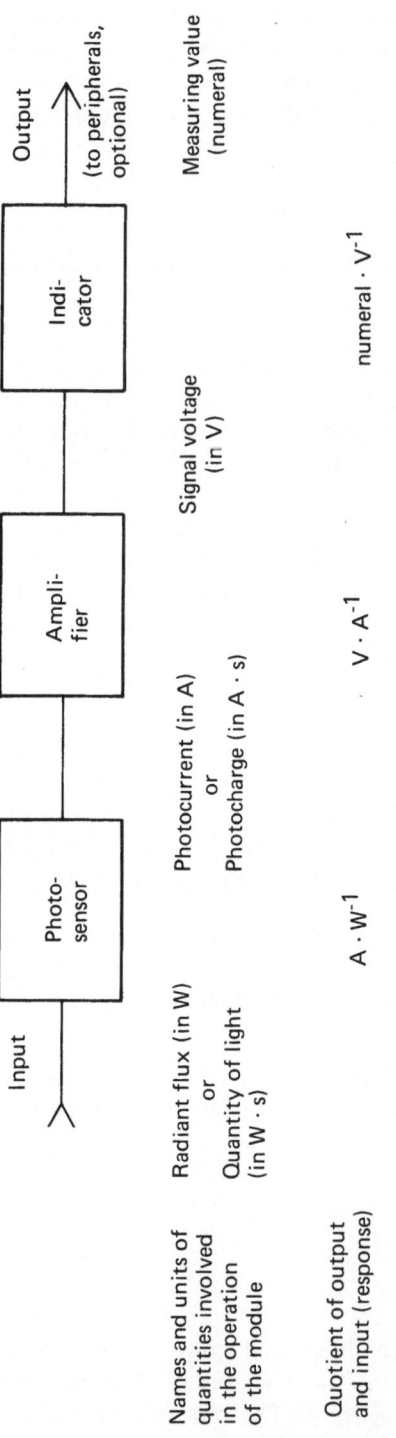

Input	Photo-sensor	Ampli-fier	Indi-cator	Output
				(to peripherals, optional)

Names and units of quantities involved in the operation of the module

| Radiant flux (in W) or Quantity of light (in W · s) | Photocurrent (in A) or Photocharge (in A · s) | Signal voltage (in V) | Measuring value (numeral) |

Quotient of output and input (response)

$$A \cdot W^{-1} \qquad V \cdot A^{-1} \qquad numeral \cdot V^{-1}$$

Fig. 7.4. Basic photoelectric assembly. *Function:* During irradiation the photosensor supplies an electric current (or a charge) to the amplifier. This converts and amplifies the input quantity into voltage and then supplies this to the indicator. The latter produces the measuring value; this may be displayed either on a scale (analogue) or in numbers (digital), and can be supplied to peripheral units such as a printer, etc.

Notes:

1. In quantities, where time is involved, the photometric system is responding to energy instead of power.

2. The quotient of measuring value and radiant flux impinging on the photocathode or that of measuring value and quantity of light impinging on the photocathode, determines the sensitivity of the whole photoelectric assembly

of modulated light have already been described in Chapter 4, Section 5 so that we now have to point out the advantage of operation with direct light. This advantage consists in the short sampling time that is achieved with direct light and short sampling times are required in high-speed scanning procedures (area-or wavelength-scanning).

The *sampling time* is the time that passes between the moment at which the photocathode is illuminated or the illumination of the photocathode is changed to the moment at which the full measuring value corresponding to the full radiant flux or the full change in radiant flux is displayed. With direct light, theoretically, the shortest possible sampling time is the sum of the electron-transit time and of the rise time in the photomultiplier tube; this is about 10^{-7}s. In practice, however, this is not achieved, because of additional time required for the operation of the rest of modules linked to the photomultiplier tube; the actual minimum time being in the order of magnitude of microseconds.

On the other hand, with a short sampling time the effect of irregular statistical fluctuations, e.g. noise is much greater than with a longer time. In practice a compromise has to be reached between a short sampling time and the suppression of disturbing noise.

The shortest sampling times are obtained with purely electronic indicating systems such as an oscilloscope or light-emitting diodes that are incorporated in digital devices (digital voltmeters) including desk calculators and computers. These systems also permit the averaging and display of the mean of a certain number of measuring values, with the result that the effective sampling time is a multiple of the time of a single value of the averaged series. The averaging is achieved with special electronic hardware, or else by means of a desk calculator or computer in on-line operation. Of course, the property of the stage object to be measured must not vary and there must not be any drift in the measuring value during the time of averaging.

Galvanometers of standard design or pen recorders allow only moderate sampling times; down to one tenth of a second. With these the sampling time can be extended by damping, hence they are mostly equipped with means for continuous or stepwise variation of the strength of damping. The effect of damping is shown in Figure 7.5.

Variation in Sensitivity of the Photoelectric Assembly. The power supply for the photomultiplier and/or the response of the interface (amplification) should be regulatable, both continuously and stepwise. This is in order to supply a sufficiently large voltage to the indicator and to compensate the extremely large variations in radiant flux incident on the photocathode, that can be expected with photometric work with the microscope, or the variation of spectral sensitivity in the photocathode. For each set of comparison measurements continuous regulation allows the measuring value for the reference material to be exactly set to the true value of its interaction factor, e.g. 100% transmittance or true reference-reflectance with the consequence that the indicator immediately indicates the true value of interaction factor of the specimen. In a set of measurements stepwise regulation allows the measuring values to be kept in the same order of magnitude and a similar precision in the mathemati-

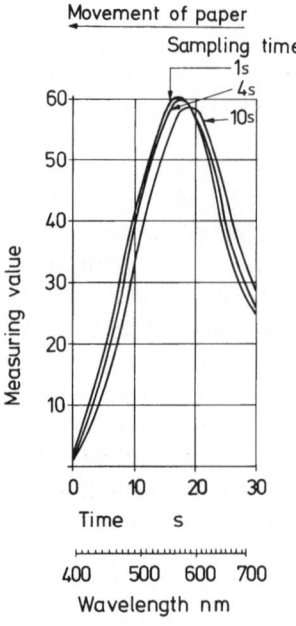

Fig. 7.5. Recorded variations of measuring values for various sampling times. In this test the spectral variation of the measuring value from the photosensor was recorded on a SERVOGOR recorder for synchronous movement of the recording paper and wavelength drive of a monochromator. The time to traverse the spectral range 400–700 nm was about 30 s, and the paper travel was made 6 cm for this time. The pen was adjusted first to take 1 s to reach its maximum amplitude (measuring value $100 \cong 20$ cm height of paper), this being the undamped sampling time.

Next the pen was damped for a sampling time of 4 s and a new curve recorded; this was repeated with a sampling time of 10 s. It can be seen that, for any wavelength, there is a lag when the pen is damped, and this lag increases with the sampling time. The lag results in a displacement of the peak inverse to the direction of the movement of the paper, hence towards longer wavelengths. It can be seen also that, in both the 4 s- and the 10 s-curves, the peak is lower; this is due to the pen receiving lowered measuring values before it has had time to record the maximum

cal processing of the values to be obtained. Of course, the setting of power supply and amplification must be reproducible, while the factors by which the change in power supply or amplification affect the measuring value must be known (calibrated amplifier).

7.3.2 Linearity in Conversion

Linearity Tolerance. Electric indicating devices are divided in classes according to their linearity tolerance (German Standard VDE 0410). Perfect linearity cannot be obtained because of fluctuation due to internal currents and the non-perfect behaviour of the electric modules. The tolerance is expressed as the difference between the indicated value and the true value; this can be expressed (usually in percent) in either of two ways.

1. It can be normalised to the maximum reading of the scale in which case the error is the same over the whole scale. This normalisation specifies the category; for category no. 1 the tolerance is 1% of the maximum reading on the scale.

2. It can be normalised to the true value, in which case the error increases to infinity as the true reading decreases to zero. If the true reading is one-tenth of the maximum then for the same category the error tolerance would be 10% of the true reading. For this reason, the upper two-thirds only of the scale should be used, if possible.

Note: Attention should be paid to the ambient temperature because an indicating device is calibrated to work within a specified range of temperature; the apparatus should always be given time to achieve a temperature balance before readings are taken.

Checking of Linearity. When a microscope photometer is delivered to the pur-
chaser all electrical modules comprising it will have a linear response with
tolerances of nonlinearity according to the previous paragraphs. But as time
passes the linearity may become affected, and so it is desirable to check this
from time to time; the basis of the procedure is to change the radiant flux
of the entering light by a known amount and observe whether the measuring
value changes in the same proportion. There are two kinds of auxiliary apparatus
that can be used for this purpose.

1. Absorbing filters of known transmittance, or else a calibrated double
wedge which gives a range of transmittance; these are placed anywhere in
the optical train, preferably immediately in front of the photocathode.

2. A triplet of polars, the middle one of which is rotatable between two
(fixed) outer ones, the vibration directions of which are parallel to each other.
The transmittance of the whole triplet is a function of $\sin^4 \alpha$ where α is the
angle of rotation of the central polar from the position of parallelism with
the two (fixed) outer ones. Thus the expected change of reading can be predicted
from the rotation angle of the inner polar and then compared with the experimental reading.

Note: Intensity can, theoretically, be controlled by a pair of polars, but then the transmitted
vibration of one of the two has to be varied, which would not maintain exact similarity of working
conditions for the purpose being discussed.

The tests are repeated at different overall sensitivities of the apparatus and
this is conveniently done simply by changing the amplification of the photocurrent. No special adjustment of the apparatus is required, but the response
should be such that, on interchange of the test device, the signal moves across
the range of reading values that is of interest for the subsequent use of the
apparatus.

Drift in the Measuring Value can be caused by:
1. Drift in electric power supply by electric modules; this is mainly a function
of temperature and can be minimised by operation at constant temperature.

2. Ageing of the lamp, which is cured by not running a lamp too long
before replacing it.

3. Fatigue in electric circuits or in the photosensor can again be cured
by earlier replacement by new parts.

The effect of drift can be much reduced by interchanging specimen and
reference material at each wavelength, rather than running one for each wavelength and then repeating with the other. Another way of achieving this reduction
is to use a double beam photometric system.

CHAPTER 8

Adjustment of Equipment

The conditions of adjustment have already been described for parts of the equipment, so that references only will be given to these in the present chapter. Likewise the errors in measurements arising from imperfect adjustment are described in Chapter 11, so that they too are merely referred to here. We assume that the equipment is in rough adjustment and we set out the steps that should be taken to put it into perfect adjustment.

8.1 Microscope Axis and Stage-Rotation Axis

With rotatable stages a centring device is required to move the image of the rotation centre of the stage into the centre of the ocular diaphragm (or photometer diaphragm). The centre of this diaphragm may be marked by a pattern, such as cross-wires (reticule), or a small circle. The centring device can be attached to the objective holder, the objective mount, or the stage. But correct centring is achieved if the centring device of only one of the three is operated, the others having pre-set and fixed adjustments. Objectives are supposed to be parcentric, but a slight correction of centring at each exchange of objectives may be required.

8.2 Luminous-Field Diaphragm

1. Look at the stage object and reduce the diameter of the luminous-field diaphragm until it is seen well inside the limit of the field.

2. Form a sharp image of this diaphragm (Fig. 8.1 a). This is done in transmitted light by focusing the condenser; in reflected light it is done by moving the diaphragm or a lens between the diaphragm and the reflector along the axis of the illuminator.

3. Centre the diaphragm (Fig. 8.1 b) and then open it out, until the whole field is just clear (Fig. 8.1 c).

Centring should be available either on the diaphragm itself or else on the lens that images it on to the stage; one of these has to be pre-set by the manufacturer or else the observer cannot centre the diaphragm.

Notes:

1. If only the central part of the field is properly illuminated, as can happen with a low-power objective, the condenser is of too short a focal length and should be replaced by one of longer focal length (Chap. 6, Sect. 1.1).

2. Adjustment for photometry is discussed in Chapter 11, Section 3.

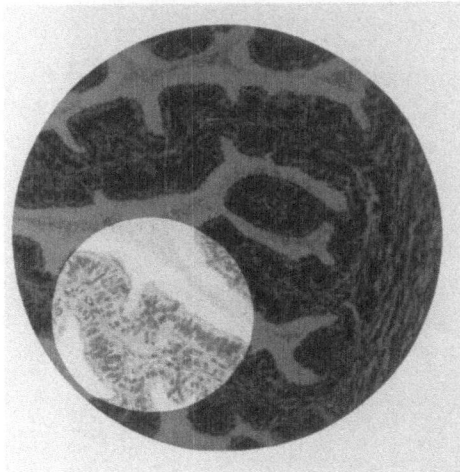

Fig. 8.1 a–c. Image of luminous-field diaphragm formed by the condenser in the stage object.

(a) Condenser (transmitted or reflected light) correctly focused so that the image is sharp. (b) If the image is not centred the diaphragm should be centred, if the diaphragm is not provided with such an adjustment, the manufacturer should be consulted. (c) For viewing, the diaphragm should be opened until it coincides with the edge of the field. (For adjustment for photometric work see Fig. 11.3)

(a)

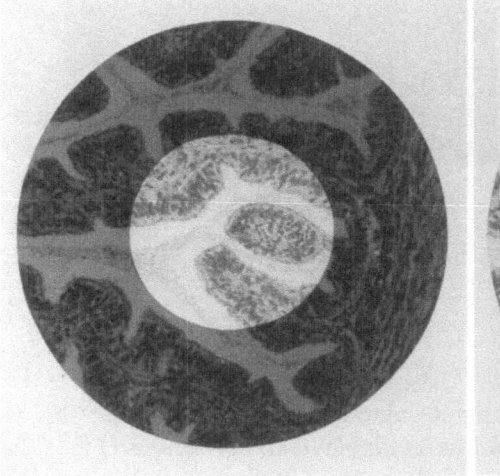
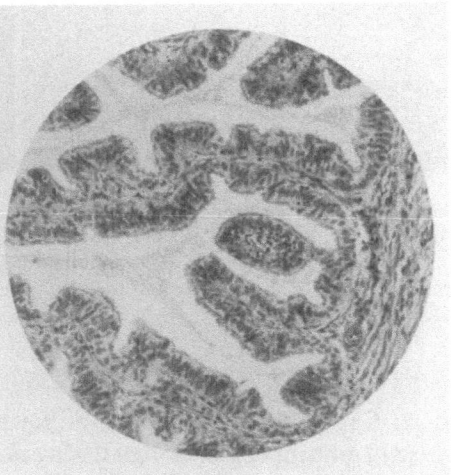

(b)   (c)

8.3 Condenser-Aperture Diaphragm and the Luminous Surface of Lamps

1. Insert the Bertrand lens (Chap. 2, Sect. 3.2) and reduce the diameter of the condensor aperture diaphragm to one-half to two-third of the aperture area of the objective.

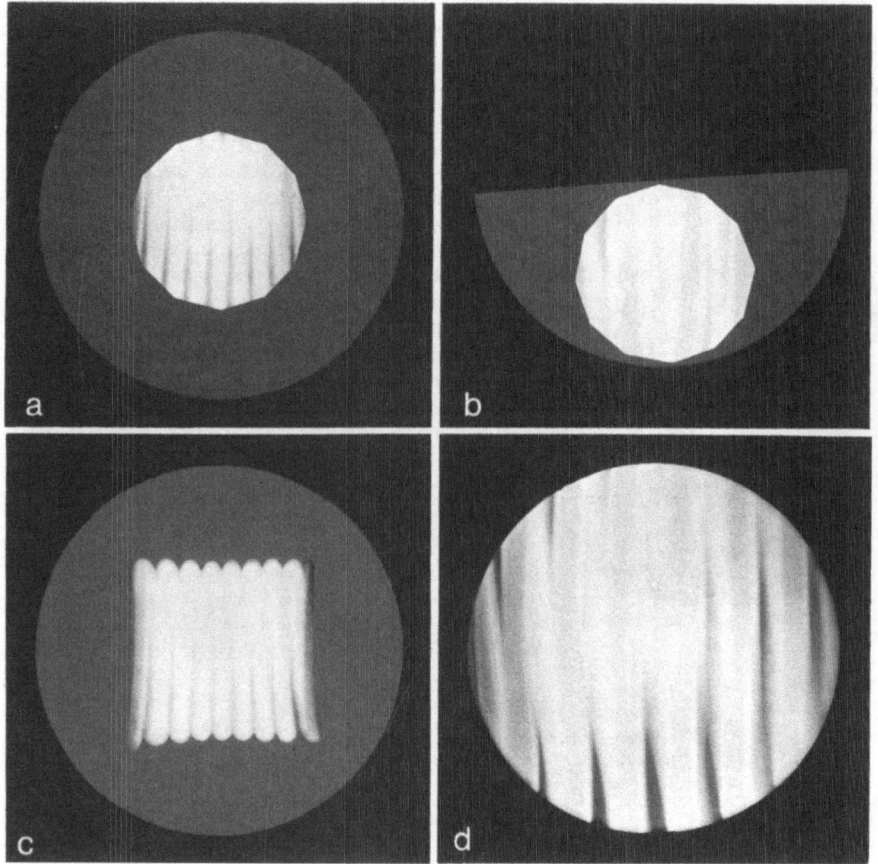

Fig. 8.2a–d. Various adjustments of the condenser-aperture diaphragm and lamp filament. The figure shows photographs of the aperture area of the objective.
(a) Diaphragm centred. (b) Diaphragm adjusted for illumination with a prism reflector. (c) Image of lamp filament centred. (d) Image of lamp filament enlarged in order to fill the aperture area

2. Centre the diaphragm by moving it sideways in its own plane; this is a useful adjustment, although it may not be available in all condensers (Fig. 8.2a and b).

3. Move either the lamp or the collector along the illumination axis until the image of the filament (or of the discharge arc) is as sharp as possible; any ground-glass filter should be removed from in front of the lamp for this step.

4. Centre this image by displacing either the lamp or the collector sideways; one of these has to be pre-set by the manufacturer or else the operator cannot centre the image properly. (Fig. 8.2c and b).

5. Check whether the field is uniformly illuminated and, if required, improve the illumination by altering the distance between the lamp and the collector.

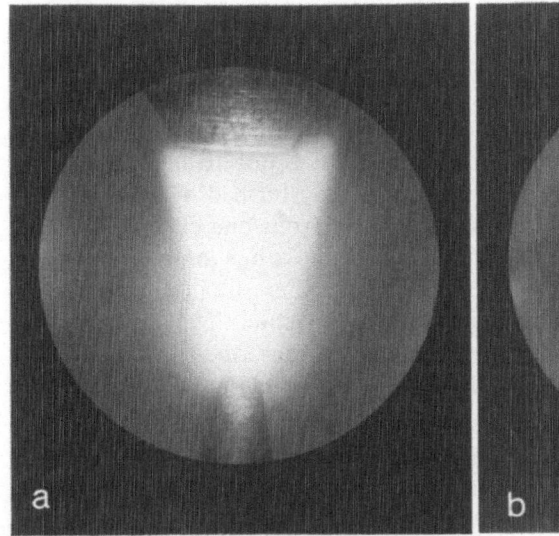
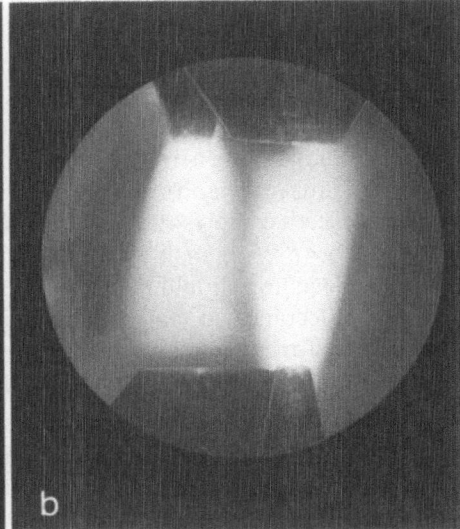
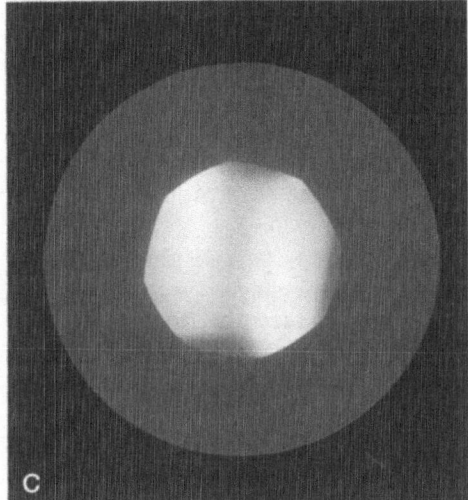

Fig. 8.3a–c. Correct adjustment of the arc of a discharge lamp shown with an electrode shape as in the XBO 150 model viewed in the aperture area of the objective.

(a) Centred image of discharge arc. (b) Discharge arc and image of the back of this. The discharge arc and the concave mirror that forms an image of the back of the arc have to be adjusted so that the image of the back side lies by the side of the arc and that the arc together with its image fills the centre of the aperture area of the condenser. (c) Diameter of the condenser-aperture diaphragm reduced so that the diaphragm just contains the most efficient portion of the front and back of the luminous surface

Notes:

1. In reflected light the image of the condenser-aperture diaphragm is reflected by the stage object; it is, therefore, necessary to ensure at this stage that the stage object is level (see Sect. 4).

2. In reflected light with the prism reflector the diaphragm has to be displaced from the centred position in order to supply light only to the half of the prism that is operative (Fig. 8.2b) unless the prism is pre-adjusted by the manufacturer so that when the prism is inserted, light is automatically supplied to the appropiate half of the prism.

3. If illumination of the total objective-aperture area is required but the image of the filament or discharge arc is too small (Fig. 8.2c), some enlargement can be obtained by altering the distance between the lamp and the collector without noticeably reducing the sharpness of the image of the filament (Fig. 8.2d). If this is insufficient, then a collector of shorter focal length must be used; alternatively a lamp with a larger luminous surface could be substituted. Enlargement can also be obtained indirectly, i.e. by forming an image of the back of the luminous surface in the plane of the front side with the help of a concave mirror. This procedure is mostly applied with discharge-arc lamps (Fig. 8.3b).

8.4 Levelling of the Stage Object in Reflected Light

Perfect levelling is, of course, possible only if the microscope-stage surface is perfectly normal to the microscope axis; for this the manufacturer has to be responsible. For general observation it may suffice just to level the whole polished mount in a hand-press. However, if the image formed by a high-power flat-field objective shows deterioration of sharpness from one edge of the field to the opposite edge or becomes unsharp when the stage object is moved, then better levelling should be carried out. This is quite essential when measurements of reflectance at normal incidence are to be made. The facilities that may be used for perfect and convenient levelling are described in Chapter 11, Section 2. In the following we are dealing with the checking of the levelling.

Levelling on a Non-Rotating Stage. The levelling can be checked by observing whether the image of a mark in the centre of the objective-aperture area, that is formed after reflection at the stage object, exactly coincides with the original mark or not (autocollimation method). We use as the mark the image of the condenser-aperture diaphragm of which the diameter is reduced as far as possible. This image must be precentred by using a glass-plate reflector together with a very low-power objective and an ocular with centred cross-wires so that the image is in the centre of the cross-wires. We observe the image with the Bertrand lens when it is formed after reflection at the surface of a specularly reflecting mirror that has plan-parallel faces so that the surface is parallel with the surface of the stage. We proceed in the following way:

1. Fix the front-surface mirror on the microscope stage so that its whole lower surface is in close contact with the surface of the stage.

2. Focus the objective on the mirror surface, reduce the diameter of the luminous-field diaphragm, and look at the aperture area of the objective and the image of the condenser-aperture diaphragm with the help of the Bertrand lens.

3. Reduce the diameter of the condenser-aperture diaphragm to the smallest possible size and adjust this diaphragm so that the centre of its image coincides with the centre of the ocular cross-wires, and fix the diaphragm in this position.

4. Focus the objective under the same conditions of adjustment of diaphragms as described above on the actual stage object.

5. Look at the aperture area of the objective and check whether the image of the condenser-aperture diaphragm is displaced from its centred position or not.

6. If any displacement is observed, adjust the levelling of the stage object so that the centred position of the image of the condenser-aperture diaphragm is restored.

Levelling on a Rotating Stage. If a rotating microscope stage is used, the levelling can be checked in an easier way: simply by observing, whether with the stage object in focus, the image of the condenser-aperture diaphragm moves during the rotation of the stage. When it moves the levelling is not perfect.

Note: The lower the magnifying power of the objective the greater will be the displacement of the image of the condenser-aperture diaphragm from its central position in the aperture area for a given tilt of the reflecting stage-object surface. If the central distances in the aperture area have been calibrated in terms of angles subtended between the microscope axis and rays intersecting the axis in the plane of the stage object according to Appendix 14 the angle of tilt can be measured.

8.5 Crossing the Polars and Orienting the Cross-Wires

The polars are described in Chapter 5, Section 6. The polariser is oriented in the standard way with its permitted vibration direction ($0°$ —rotation position) East-West (Fig. 5.8); this is normal to the plane of incidence of the reflector in reflected light. We start with the description of the adjustment in transmitted light, since the procedure is simpler than in reflected light. In transmitted-light studies the exact orientation of the vibration of the polars is not important, and East-West and North-South orientation is always used just because the eye is uneasy otherwise, nor is the required accuracy of crossing of the polars very high. But it is important to have the ocular cross-wires parallel to the vibration directions of the crossed polars, and thus to use the cross-wires to indicate the vibration directions. It is otherwise in reflected light where the polars have to be set symmetrically with respect to the plane of incidence of the reflector, which intersects the stage along its North-South line; also, the required accuracy of crossing is higher.

Notes:

1. If the microscope is to be used for both reflected and transmitted light, the polars should be set in reflected light because the setting has to be more accurate than for transmitted light.

2. For photometric work the analyser is withdrawn except in interference systems such as in Figure 10.1, and there must be no polarisation effects in mirrors or beam splitters situated in the imaging optical train. If such effects cannot be avoided they have to be compensated with the help of a plate producing circularly polarised light in the imaging optical train.

Transmitted Light. 1. If the polariser does not have its vibration direction engraved, this must first be ascertained by crossing it with a polar of known vibration direction (a piece of polarising foil can be used for this); this direction is then set East-West.

2. Against a white background, by turning the ocular the cross-wires are set East-West and North-South, which can be done accurately by the eye alone.

3. We select a small birefringent crystal that has a morphological direction parallel to one of its principal vibrations; any other material showing such a direction (foil or fibre) can be used instead. This is set with its direction along one of the cross-wires as exactly as possible. A strain-free objective of the lowest power should be used so as to have as much light as can be provided by the optical system.

4. The analyser is now inserted with its vibration direction North-South; the test should be made in orthoscopic viewing and the Figure 8.4 should be consulted for the results of the test. The required adjustments are then completed.

 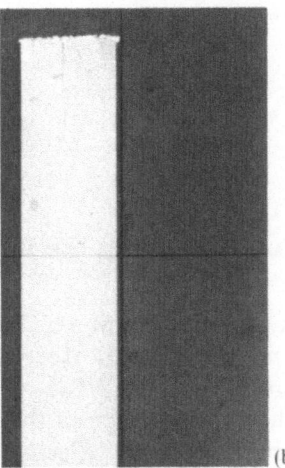

(a)   (b)

Fig. 8.4a and b. Image of a birefringent test object in transmitted light. A sharp and long edge of the test object (parallel to a principal vibration in the test object) is set parallel to one of the cross-wires. The crossing of polars results in perfect extinction in the field outside an image of the test object.

(a) Perfect extinction also in the test object indicates that the two vibrations in the test object are parallel to those of the polars and hence the two vibrations of the crossed polars are parallel to the cross-wires. (b) Brightening in the test object indicates that the cross-wires are not parallel to the polars. Either the test object together with the cross-wires or both polars together must be rotated until the situation as in (a) is achieved

Note: If the analyser and cross-wires have previously been set in reflected light, but the microscope is now to be used in transmitted light, with the analyser in place, the transmitted-light polariser is inserted and rotated until perfect extinction is obtained for the empty field in orthoscopic viewing.

Reflected Light. The viewing of the aperture area of the objective through the Bertrand lens is recommended for checking the setting of polars (conoscopic viewing).

A homogeneous specularly reflecting substance, e.g. a reflectance standard (Chap. 9, Sect. 3.2) or a front face mirror is levelled and inspected using a glass-plate reflector so as to have the whole aperture area in view, for which purpose also a high power objective is used; an oil-immersion objective of high numerical aperture is best because its numerical aperture can be higher than that of any dry objective. The condenser-aperture diaphragm should be opened, the diameter of the luminous-field diaphragm is somewhat reduced.

1. The polariser for reflected light is inserted with its vibration direction set East-West as accurately as possible at this step. There are two results of this test (Fig. 8.5) and the figure should be consulted for these, after which the polars are accurately crossed and oriented.

2. It remains to see that the cross-wires lie exactly along the arms of the black cross of Figure 8.5b.

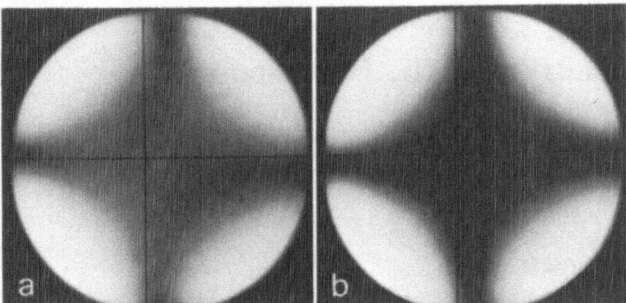

Fig. 8.5a and b. Extinction cross viewed between crossed polariser and analyser in the aperture area of the objective. A high-power objective is used in order to obtain a large aperture angle. The stage object is a specularly-reflecting isotropic substance.

When the centre of the cross is not extinguished but is neutral grey (a) the polariser is correctly oriented, but the analyser is not perfectly crossed to it; we turn the analyser until perfect extinction is obtained and orientate the ocular cross-wires along the arms of the cross (b).

Note: When the centre of the cross is coloured, the polars may be crossed or not but the polariser is not normal to the plane of incidence of the reflector. The polars should be turned together, by trial and error, until perfect extinction is achieved

CHAPTER 9

Procedures in Photometry with the Microscope

9.1 Principle of Measurements

Photometry with the microscope is both a special use of photometry and a special use of the microscope. The present chapter is concerned with the latter, and some knowledge of its contents is required for proper use of the microscope for this purpose; otherwise the results tend to be of poor quality. The definitions of technical terms and the nature of the quantities involved are discussed in Appendix 2. For more information about photometry in general the following literature may be consulted: Walsh (1953), Ewing (1954), Kortüm (1962), Reeb (1962), Wendtland (1968), Weissberger and Rossiter (1972).

The method used is not an absolute measure of light power (radiant flux), but a comparative one which involves a previously calibrated reference material. It can be called a method of consecutive comparison; it involves obtaining two correct measuring values (G) and relating the one for the specimen to the one for the reference material. We shall use the symbol IF (interaction factor) as this can apply to the measurement of any property of the stage object that is proportional to light power, either in transmitted- or in reflected-light, and the simplest relation is

$$\frac{IF_{sp}}{IF_{ref}} = \frac{G_{sp}}{G_{ref}} = q \tag{9.1}$$

G	measuring value (Galvanometer reading, indicated or displayed value) (numeral)
sp	specimen
ref	reference material (standard)
q	expression of the ratio (see Fig. A.3)

In the special case of absorbance measurements (Sect. 5.3) a logarithmic scale is used and the quotient is inverted

$$\lg\left(\frac{IF_{ref}}{IF_{sp}}\right) = \lg\left(\frac{G_{ref}}{G_{sp}}\right). \tag{9.2}$$

Note: For the relation between interaction factor and radiant flux see Eqs. (3.1) and (3.4).

9.2 Photometer Systems

The operational background of the three main types of photometer systems is described in this section: description of technical design is omitted. All three types have the same basic modules; the microscope, the photoelectric assembly and between them the photometer head containing the photometer diaphragm, but they differ in optical geometry of the photometric reference beam.

Type no. I is the *single-optics system with stage-object switching* (Fig. 9.1a). In this system the geometry of the photometric beam including that of the illuminating beam must remain unchanged when specimen and reference material are interchanged. For the interchange the specimen and the reference material must be bodily moved, and in order to maintain the geometry of the optical train the specimen and the reference material must be placed in the same plane of focus and must be level (Chap. 11, Sect. 2). The operation of the lamp and photometric assembly must be stable at least for the period of measurements on the reference material and on the specimen (or set of specimens) that is to be compared with the same reference material.

Type no. I has the advantage of (1) being conveniently attached to all types of microscope; (2) having equal optical path and transmittance in the light train for both specimen and reference material [see Eq. (3.4)]; (3) being less costly than other types.

Type no. II is the *single-optics system with beam switching* (Fig. 9.1b). In this system the geometry of the photometric beam in front of the photocathode remains almost unchanged when the measuring spot in the image field or in the stage-object field is switched from the specimen to the reference material, but the direction of the photometric beam is changed. The specimen and the reference material must be sufficiently small for them to lie together in the full image field of the microscope; the photometric beam is switched so that the photocathode receives light first from one and then from the other. This can be done in three ways, as described in Figure 9.1b.

This type does not allow the luminous field to be adjusted individually and properly in respect to the photometric field, hence it allows glare (Chap. 12, Sect. 3.1). The optical trains for the measurement of the specimen and of the reference material are not strictly equal, and systematic errors may occur according to the $\cos^4$-law (Chap. 12, Sect. 3.3). But these systems have the following advantages:

1. High speed of comparison of pairs of measuring values from different spots on the stage object (more than 50 pairs of measurements can be made in one second), and hence no high demand on the operational stability of electric modules.

2. Immediate display of variation of comparison results with wavelength (absolute spectral curves), or with environmental conditions, e.g., temperature.

Type no. III is the *double-optics system* (Fig. 9.1c). In this system, in addition to the optical train of the microscope, there is another, called the control train. The latter supplies the radiant flux needed as reference and this is first calibrated by the radiant flux from an actual reference material. The original train is split off at any point between the monochromatising device and the

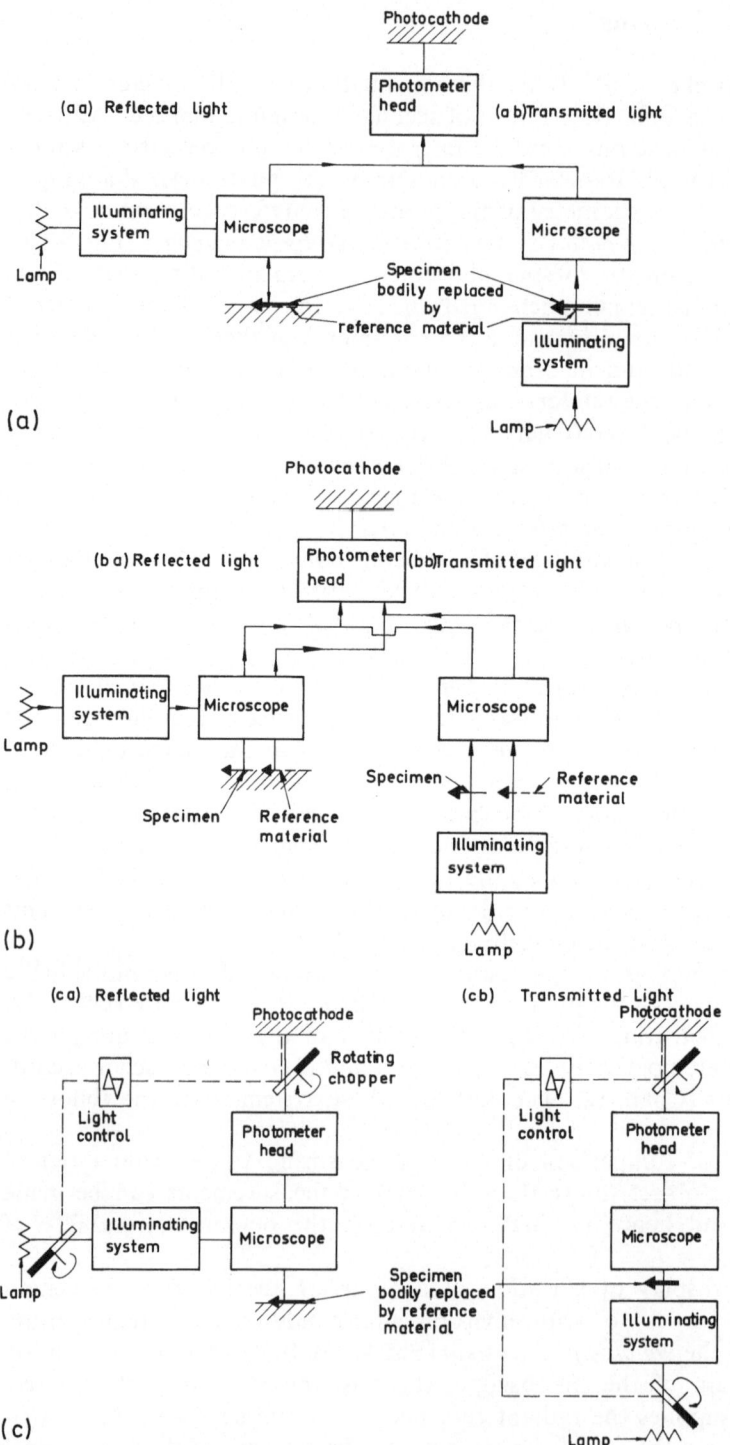

Fig. 9.1 a – c. Legend see opposite page

◀ Fig. 9.1 a–c. Main types of microscope photometer, schematic. Each figure represents a certain type, of which many variations in technical design exist. *Full straight lines:* path of photometric train; *dashed lines:* that of control train.

(a) Single-optics system with stage-object switching (*aa*) for reflected light, (*ab*) for transmitted light. The photometric beam is selected by the photometer diaphragm and the luminous-field diaphragm, the size of which is adapted to that of the photometer diaphragm (Chap. 11, Sect. 3). Measurement is made on the specimen under given conditions of adjustment of the whole equipment. The specimen is bodily replaced by the reference material, or vice versa and the measurement repeated under exactly the same conditions.

(b) Single-optics system with beam switching (*ba*) for reflected light, (*bb*) for transmitted light. The specimen and the reference material are adjacent stage objects (in the same preparation) and the photometric beam is switched from one to the other. Thus the photocathode receives light alternatively from two different spots in the preparation.

There are three switching mechanisms: (1) With one photometer diaphragm that moves reciprocally in a line normal to the microscope axis. (2) With two photometer diaphragms, one selecting the spot for the reference material, both being alternately opened and closed by a chopper mechanism. (3) A single photometer diaphragm remains steady and the image of the specimen and of the reference material move in a line normal to the microscope axis, so that the diaphragm is illuminated first by the specimen and then by the reference material or vice versa.

The switching module is the extra item in the equipment for this type of system. For the first and second kind of switching the module is placed in the plane of the photometer diaphragm so as to make it move from one spot to the other or to make the two diaphragms operate in phase with the sampling process of the electric assembly. For the third kind of switching either a prism or a mirror is placed exactly in the aperture area of the objective and the movement is synchronised with the sampling process. Alternatively the switching mechanism can be placed in the plane of an intermediate image of the aperture area.

Note: The first and second kind of switching requires precise alignment of the photocathode in a plane conjugate to the aperture area of the objective in order to keep the illuminated area in the cathode constant, although the photometric beam is made to traverse different spots in the field.

(c) Double-optics system; (*ca*) for reflected light, (*cb*) for transmitted light. The photocathode is alternately illuminated by the regular photometric beam (*full line*) and by the control beam (*dashed line*). The alternation of light is achieved with a pair of choppers equipped with mirrors operating in phase. The figure shows the choppers in such a position that the regular photometric beam is effective.

The measuring value for the control beam serves as reference. For this purpose, with the equipment adjusted and before starting the measurement of the specimen the measuring value is equalised for the two beams by using an actual reference material in the photometric beam. The equalisation is achieved in two ways: (1) By adapting the radiant flux in the control beam to that in the photometric beam with the help of light controls situated in the control beam and acting by absorption or reflection. When the radiant flux in the two beams is equal also the measuring values are equal. (2) By adapting directly the measuring value for the control beam to that for the photometric beam with the help of electronics, so that no light controls are needed.

The measuring value for the control beam being equalised in the described way remains or is kept constant as long as the adjustment of equipment is not changed and is used as reference for the subsequent measurements with the specimen in the photometric beam

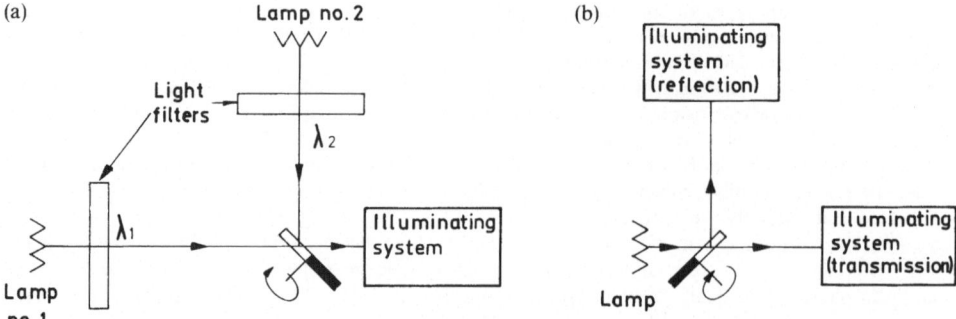

Fig. 9.2a and b. Illuminating-beam switching device.
   (a) The illumination is supplied by two lamps, each of which is combined with a monochromatising and an intensity controlling device. One of the monochromatising devices is set for wavelength
λ_1 the other for wavelength λ_2; the intensity control allows to balance the spectral response of
measuring equipment for the wavelengths λ_1 and λ_2. The illuminating beam is switched from
one lamp to the other. Thus the stage object receives light alternately with two different wavelengths
and the photoelectric assembly is controlled so that the ratio of the two measuring values for
the wavelength λ_1 and λ_2 is immediately displayed or processed. (b) The illuminating beam is
supplied by a single lamp but switched from the illuminating system for transmitted light to
that for reflected light. Thus the stage object receives light alternately from two different directions
or, in general, with two different kinds of illumination and the photoelectric assembly is controlled
so that the ratio of the measuring values for two different kinds of illumination is immediately
displayed or processed

condenser and is then recombined with the control train immediately in front
of the photocathode.

   There is a stringent condition about equality of optical path and transmittance in the two trains, unless certain errors, due to inequal light transport
through both trains are tolerated or else compensated optically and/or electrically. Perfect equality requires a twin microscope allowing synchronous change
of lenses and adjustment of diaphragms, but for moderate demands on accuracy
or for operation with optical or electrical compensating devices, the control
train can be equipped with simple lenses or fibre optics (Smith, 1975), these
being much cheaper and easier to manipulate than the twin microscope.

   Double-optics systems have the following advantages:
   1. High speed in comparison of pairs or series of measuring values.
   2. Immediate display of absolute spectral curves.
   3. No restriction in size and position of either specimen or reference material.
   4. No high demand on the stability of electrical conditions.
   5. Greatest versability in respect to combination with a computer or other
kind of automation device (Sect. 6).

Note: With a special device the illuminating beam can be switched from one lamp to another
or from one illuminating system to another. The function of such a device is shown in Figure 9.2,
where it is seen that it operates like a light-modulating system; in the switching system the light
is not modulated so that either light or no light is alternatingly supplied but so that the nature
of two light-supplying systems is changed alternately. Examples for application of this kind
of photometric comparison have not yet been published but may, in future, become important.

9.3 Reference Materials

9.3.1 Transmission Reference Material

The simplest material is an empty spot in the embedding material of the specimen. The reference-measuring value is taken as unity (or 100%) so that the measuring value for the specimen immediately indicates the true value of transmittance. (For measurements of intensities due to interference as described in Chapter 10, Section 2, the reference material must have an optical thickness that does not exceed a difference of half a wavelength with the optical thickness of the specimen.) When the transmittance is to be converted into absorbance the empty spot must indicate absorbance zero.

Note: For "pattern recognition" it is sometimes preferable to normalise the readings to the value for a certain feature of the specimen considered as a standard; in any case, measuring results are always relative.

Reference materials of macro size are most suitable for checking the linearity of the measuring equipment or for establishing the power of the spectral resolution or of the resolution in brightness differences. Among such reference materials are neutral filters or wedges, didymium glass (Fig. 11.6), and special liquids that have reproducible and predictable absorption.

9.3.2 Specular-Reflection Reference Material

For work in reflected light stringent demands are made on the material and on the condition of its surface. This surface must
  1. be extremely hard so as not to be easily scratched;
  2. be extremely resistant to chemical change;
  3. be isotropic;
  4. be perfectly flat and polished, with a flawless area of at least 4 mm^2 because the calibration is macroscopic.
  5. Have a fairly flat curve of spectral reflectance over the wavelength range to be used.
  The following substances are accepted by the international bodies concerned and available from manufacturers of microscope photometers; items 1, 2, and 3 by the Commission on Ore Microscopy of the International Mineralogical Association and 4 and 5 by the International Commission on Coal Petrology (International Vocabulary of Coal Petrology, 1971).
  1. Dark neutral glass (Type NG 1, Schott, Mainz, FRG) with a mean reflectance in air of 4.5%.
  2. Single crystal of black variety of silicon carbide (SiC) with the basal face (0001) polished; mean reflectance in air 20% and in standard immersion oil 7.5%.
  3. Single mixed crystal of (W, Ti) C with the basal face (0001) polished; mean spectral reflectance in air 47% and in oil 33%.

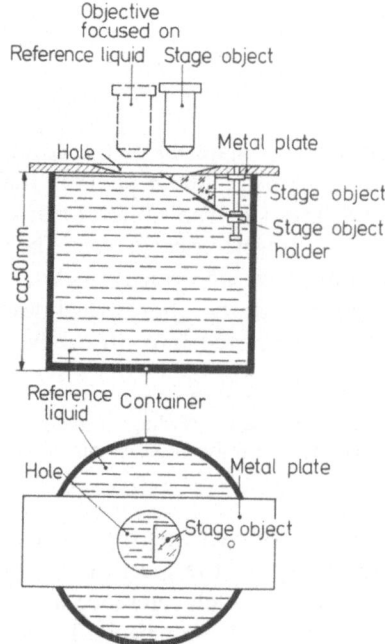

Fig. 9.3. Surface of a liquid used as reference for the measurement of specular reflectance in air. The liquid e.g. water, immersion oil, etc. must be clear in the sense that it does not scatter light; it is placed in a container having black surfaces inside to suppress internal reflections. The reflectance of the liquid at normal incidence is calculated with the help of the formula

$$R = \left(\frac{n-1}{n+1}\right)^2 \tag{9.3}$$

n refractive index of the liquid

The figure shows, as an example, that a specimen can be dipped in the liquid, fixed at a metal plate and orientated so that its surface is parallel to the surface of the liquid. In this case the reflectances of the liquid and the specimen can most conveniently be compared. The specimen is taken to be a prism.

Note: The use of a liquid as reference material requires exact vertical orientation of the microscope axis so that the axis is normal to the surface of the liquid, which is always strictly horizontal

4. Single crystal of leuco-sapphire with the basal face (0001) polished; mean spectral reflectance in air 7.5% and in oil 0.5%. This material is in the shape of a prism in order to prevent internal reflections from reaching the objective.

5. A set of glass prisms with a large cathedus-face polished and having the same shape as the sapphire; reflectances being in the range from 5% to 10% in air (see ASTM D 2798-72).

At present, calibration in oil is not done macroscopically. For substances in which any absorption is so small as to be negligible, that is having an absorption coefficient less than 0.001, e.g. SiC, the reflectance in oil is calculated according to Eq. (9.18), i.e. with the help of the measured reflectance of the substance in air and the refractive index of the oil.

Calibration in oil of stronger absorbing material, e.g. (W, Ti) C, is done microscopically using the reflectance in oil of a nonabsorbing or weakly absorbing substance as reference and operating with the lowest power oil immersion objective available.

Another reference material that can be used for measurements in air is the surface of a liquid contained in a box having black surfaces inside to absorb the internal reflections that would otherwise cause errors (Fig. 9.3). In this case, the refractive index of the liquid is used in order to calculate its reflectance in air as no error can arise on account of the nature of the surface. For the calculation Eq. (9.3) is used. This material is suited for the microscopic calibration of any surface material that has a low reflectance, as well as for the calibration of any surface material in the UV or near IR range. The refractive indexes of water for these ranges are found in table books, e.g. Landolt-Börnstein (1962). Those of immersion oil or immersion glycerine can be obtained from microscope manufacturers.

Note: Some people recommend calculating the reflectance of a transparent solid reference material from its refractive index with the help of Eq. (9.3). This must not be done because the refractive index of the border face is different from that in the interior of the substance. For example, the thickness of a border layer of a polished glass body can be up to some tenths of μm and the refractive index may be increased up to 10% (Rayleigh, 1937). Consequently, solid reflectance-reference materials should always be individually calibrated with the help of a photometer.

9.3.3 Diffuse-Reflection Reference Material

The nature of diffuse reflection is described in Chapter 10, Section 3. An ideal reference material must not produce preferential reflection in a given direction at a particular angle of incidence. Further this material must not absorb any of the light and must act in the desired way over all the spectral range for which it is to be used; but these conditions can never be fulfilled properly.

Since diffuse reflectance is widely measured macroscopically, many materials have been tried out for this purpose. The best has been found to be MgO or $BaSO_4$ either deposited from smoke on a flat solid substrate, or else pressed in powder form into a tablet. Since these have a diffuse-reflectance value of practically 100% in the visible spectrum, the reference measuring value is taken as unity; thus, the measuring value for the specimen immediately indicates its true diffuse reflectance. Accurate reflectance values of reference material are given in the literature, e.g. Kortüm (1969).

For a particular type of measurement it is often best to use a calibrated standard of the same material as the specimen; thus silver grains in auto-radiography or a sheet of standard paper in the use of reflectance in quality control in paper making.

Diffuse-reflectance standards have all a short life, and so they should be prepared afresh for each run of measurements. Instructions for preparation procedures are obtainable from the manufacturer of macroscopic photometric equipment or in papers of standardising committees (see ASTM E 259-66, E 306-66).

9.3.4 Fluorescent-Reference Material

It is hardly possible, and not necessary, to specify by a physical quantity the properties of fluorescent-reference material. This is not calibrated in terms of any concept of absolute fluorescent intensity, unless it is to be used for fundamental physical studies; instead it is specified by its nature or by the amount of a chemical substance (usually in millimole) to which the fluorescence intensity is linearly related.

Biological material containing a known amount of fluorescent substance may be used as reference, e.g. bull sperms or human leucocytes (Ruch, 1973). In this case, the substance is stained with a fluorochrome of the same kind as is being used for the specimen and both are prepared in the same way.

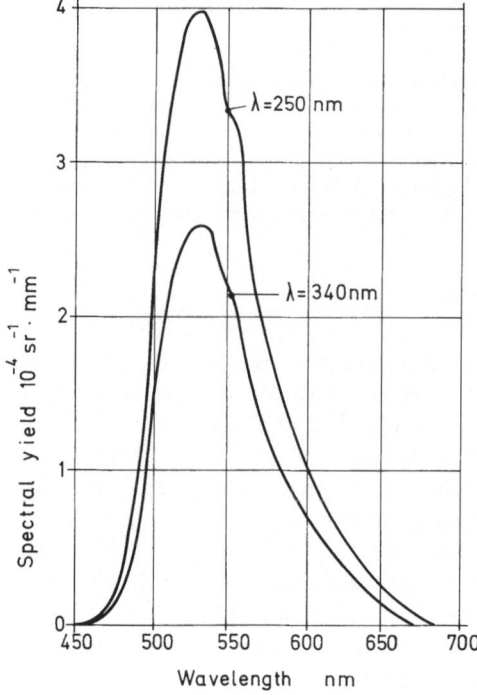

Fig. 9.4. Spectral yield of fluorescent light in an uranyl glass plate. *Type of glass:* GG 17 (Schott, Mainz, FRG), thickness 1 mm. The curves are specified by the exciting wavelengths, 250 nm and 340 nm respectively. The direction of the exciting beam is inclined at 69° to the normal of the plate.

Note: The yield is expressed as the following quotient:

$$\frac{\text{Spectral concentration of radiance of fluorescent light}}{\text{Irradiance of exciting light}}$$

$$\left(\text{in } \frac{W \cdot cm^{-2} \cdot sr^{-1} \cdot nm^{-1}}{W \cdot cm^{-2}} = sr^{-1} \cdot nm^{-1} \right).$$

Substances known as "optical brighteners" are being developed, and these may well become of importance in this connection. These have a very high yield of fluorescence and are stable over a long period.

Uranyl glass is a good fluorescent reference, and thus is prepared in small plates about 1 mm thick. These can be used as a secondary standard after calibration agianst a primary one. This glass is extremely stable under excitation, especially as compared with biological materials, which mostly tend to fade under irradiation. For uranyl glass the physical relationship between fluorescent intensity and irradiation is known (Fig. 9.4).

Also, solutions of fluorescein or fluorescein albumin in capillary cuvettes are recommended as reference materials for fluorescence (Sernetz and Thaer, 1973). Just recently, the use of uranyl and europium ion-doped glass fibres of 5–45 µm diameter have been proposed for this purpose, especially for the green and red spectral regions (Velapoldi et al., 1975). Finally, see note on photometric standard lamps in Chapter 4, Section 3.3.

9.4 Measuring Procedures

9.4.1 Measurements at a Fixed Wavelength

A series of specimens or of spots in a specimen is measured at a single fixed wavelength for some determinative purpose. In inorganic and biological material the amount of light-absorbing components (Sect. 5.3) is determined or the areal arrangement of features in a pattern is analysed. The wavelength selected is usually that for which the absorption is greatest in the material.

In reflected light such measurements are made for the identification of opaque substances; in this case the wavelength used is one that has been selected as an international standard such as 546 nm.

For the exchange of specimen and reference material, both can be moved by hand or with the help of some form of stage-object changing device. The simplest device is a mechanical stage with rectangular movements or a scanning stage (Chap. 6, Sect. 3) controlled by hand or by motors, that may be automatically operated with a computer (Sect. 6.2). This may be used in both transmitted and reflected light. For reflected-light work also special devices are available, that make the exchange most convenient (Fig. 9.5).

The measuring-value indicator may be set at unity (or 100%) for the transmittance reference or at the calibration value for the reflectance reference; otherwise it may be set to read in the top of its range for specimen or reference material, whichever has the greater value for the property to be measured.

9.4.2 Spectral Measurements on a Fixed Measuring Area

When it is desired to obtain the spectral distribution of a quantity for a given wavelength range, measurements should be made at intervals of 10 nm; a smaller interval can, of course, be used for the study of a particular feature of the spectral curve in more detail, but too large an interval can result in missing a small spectral feature (Chap. 11, Sect. 5). There are two kinds of procedure.

1. Stage-object interchange at each wavelength. This procedure requires only that the photometric system be free from fluctuations; it does not require that it be free from measurable drift for a period exceeding the time necessary for exchange of reference material and specimen. On the other hand it does make severe demands on the interchanging of the reference material and the specimen because each has to be brought back to exactly the same position for each change of wavelength; in covering the visible range (400 to 700 nm) at an interval of 10 nm some 31 such changes are involved.

The monochromatising device does have to be reset the same number of times, and the setting has to be accurately done when the interaction factor of the specimen has a great spectral variation. A set of well made notches at known wavelengths is a convenient method of rapid setting of a monochromator of the continuous-filter type. If one mistake is made in wavelength setting, only that one measurement is affected.

2. Wavelength change for both stage objects in turn (wavelength scanning). The exchange of specimen and reference is made only once in the measurement series, the spectral run being made first on one and than on the other of the stage objects. This involves two times changes of monochromator setting in equal ranges and intervals or continuous variations of wavelengths with equal speed; if one mistake is made in setting or speed, all the measurements are affected. This procedure makes heavy demands on the stability of lamp

Fig. 9.5a and b. Devices for quick exchange of specimen and reference material.

(a) Magnetic holder for reference material and quick-focusing lever (designed by the author).
1 Objective. *2* Annular plate attached to the objective. *3* Reference material in its mount fixed
in a metal ring. The ring carries magnetic screws. *4* Quick-focusing lever. *5* Auto-levelling microscope
stage. *6* Stage object.

By turning the quick-focusing lever in the direction of the arrow the microscope stage with
the specimen on it is lowered; hence the distance between the specimen surface and the objective
is increased so that the reference material can be inserted. The ring carrying the reference material
is fixed magnetically to the lower face of a ring plate in front of the objective. The reference
material is pre-levelled and brought into focus by adjusting the height of the magnetic screws,
or of the ring plate. When the reference material is taken off and the lever is turned in the
opposite direction to hit the base of the microscope stand, the focusing position of the specimen
is immediately restored.

and photoelectric assembly over a time of some minutes; only with a very steady photometric equipment it is possible to use this procedure.

Note: The two spectral curves obtained in this way merely show the spectral variation of measuring *values*, for the specimen and the reference material. In order to establish the variation of the measuring *result*, the variation of the *quotients* of measuring values must be determined and the series of quotients must be multiplied by the reference values for the corresponding wavelengths according to Eq. (9.4).

When the best possible precision is required, it is necessary to take several readings on each occasion and to average the values in order to obtain the mean measuring value and thus the mean measuring result.

The monochromatising device can be equipped with a motor for changing the setting, this being electrically controlled; the use of this avoids any observer fatigue. The most convenient operation is achieved with automatic facilities (Sect. 6.3).

9.5 Computation of Measuring Results

9.5.1 Introduction

This section deals with the measurement of transmission- and reflection-properties which are the main subjects of microscope photometry. The measurement of properties that are of minor interest is described in Chapter 10.

We must make a difference between the measuring value and the measuring result. The measuring value is directly read off the indicating device; it is given in a numeral, either corresponding to the amplitude of deflection of a pointer or mark or immediately displayed in the form of digits by light-emitting diodes. We symbolise it by G (galvanometer reading).

The measuring result is a secondary quantity, because it is determined by at least two measuring values. According to the nature of the parameters involved in the measuring result we call it direct or derived. The difference between a direct and a derived result is described in the following three sections along with some important examples of derived results. In these sections the measuring result is always the term on the left side of the equation.

◀ *Note:* The microscope stage shown in the figure is of the auto-levelling type. In this type the specimen is placed on an elastic substrate that is pressed from below so that the specimen surface touches the lower face of a plate. This plate has its lower face pre-levelled and the specimen is viewed through a diaphragm inside it. Of course, any other type of stage can be used with the exchanging device.

(b) Lanham specimen changer stage (Crystal Structures Ltd. Bottisham, Cambridge, England). *1* Slide base. *2* Moving slide. *3* Specimen holder and levelling screws. *4* Standard holder and levelling screws.

A slide is moved between two stops on either side. This slide carries two supports that are held magnetically and can be moved across the slide. One support carries the specimen, the other the reference material. For levelling, the specimen or reference surface is pressed against the lower side of three screw heads that are adjustable in height

Here we repeat the statement expressed in Eq. (3.4) that comparison of measuring results is permissible only when they have been obtained with the same apparatus and under the same conditions. An exception to this can be made if the alteration factor is known (e.g. for change in amplification of photocurrent, in size of measuring field, or of effective-aperture angle).

9.5.2 Direct Measuring Results

The simplest sort of measuring result has already be mentioned in Section 1; it is the ratio of two measuring values expressed by Eq. (9.1).

If the value of the interaction factor of the reference material is known, we can transform Eq. (9.1) and express the interaction factor of the specimen by its absolute magnitude, as in

$$IF_{sp} = \frac{G_{sp}}{G_{ref}} \cdot IF_{ref} \tag{9.4}$$

IF   interaction factor (transmittance, reflectance, fluorescent intensity, inter-
        ference function etc.)
G   measuring value (numeral)
sp   specimen
ref   reference material

Note: With a perfectly transmitting or reflecting reference material we have $IF_{ref} = 1$, hence $IF_{sp} = G_{sp}/G_{ref}$.

9.5.3 Derived Results for Chemical Compounds in Solution

The absorption behaviour of fluid or solid solutions of certain chemical compounds, especially of those in biological substances, is proportional to the path of light travelling through the substance and to the concentration of the compound in consideration. This proportionality exists if the absorption is evaluated logarithmically, the corresponding quantity being absorbance [Lambert-Beer's law, Eq. (9.7)]; absorbance is the logarithm of the reciprocal transmittance, thus

$$A \equiv \lg \frac{1}{T} = -\lg T \tag{9.5}$$

A   absorbance (numeral > 0)
T   internal transmittance (fraction)

(For distinction between internal and total transmittance see App. 2.4.)

We can consider absorbance to be derived from transmittance and so express the measuring result in terms of absorbance as follows,

$$A_{sp} \equiv \lg \frac{1}{T_{sp}} = \lg \left(\frac{G_{ref}}{G_{sp} \cdot T_{ref}} \right) = \lg G_{ref} - \lg (G_{sp} \cdot T_{ref}) \tag{9.6}$$

A absorbance
T internal transmittance (fraction)
G measuring value
sp   specimen
ref   reference material

Notes:

1. Some types of indicating device have a logarithmic scale, so that the logarithms of measuring values are immediately displayed.

2. In practice the indicating device is often set so that for the perfect transmitting reference material the logarithmic measuring value and reference value are zero $(\lg G_{ref} = \lg T_{ref} = 0)$ and hence the value of absorbance of the specimen is immediately indicated.

Lambert-Beer's law is true for a clear (non-scattering) substance, unless there is chemical reaction between various compounds or between the disolved substance and the solvent; furthermore, this law is applied only to dilute concentrations.

The proportionality factor in Lambert-Beer's law is called molar absorptivity. It is specific for the absorbing substance, and depends on the wavelength at which the absorbance is measured. Usually the wavelength for maximum absorption is used. This law is expressed by

$$A = t \cdot c \cdot \varepsilon \tag{9.7}$$

A absorbance
t path of light in the absorbing material (in cm)
(For measurements with the microscope the path is along the microscope axis, so that the thickness of the stage object is involved.)
c concentration of solved chemical compound (in $mol \cdot litre^{-1} = mmol \cdot cm^{-3}$)
ε absorptivity (in $cm^2 \cdot mmol^{-1}$)
(For definition of these quantities see App. 2.4.)

It is evident that each of the three quantities of the right side of Eq. (9.7) can be derived from the measured absorbance if the two others are known. But only the determination of absorptivity and concentration are of interest, while the thickness is either determined by the morphological characteristics of the stage object or else measured by mechanical means or by interference methods (Chap. 10, Sect. 2).

The absorptivity is determined mainly with macrophotometric equipment, values are listed in tables (e.g. Landolt-Börnstein, 1962; Burnett, 1972); only if the material is available in tiny amounts, microscope photometry may be used. Thus the main application of Lambert-Beer's law in microscope photome-

try is the determination of the concentration of absorbing compounds in microscopic specimens, or the determination of the amount of mass of compounds,
where this is closely related to concentration (see below).

In order to determine the concentration (c) we need to measure the absorbance (A) and to know two constants: molar absorptivity (ε) and thickness
(t), according to

$$c = \frac{A}{t \cdot \varepsilon}. \tag{9.8}$$

In order to determine the amount of mass we have to replace the concentration by the terms on the right side of the following equation,

$$c = \frac{m}{t \cdot F} \tag{9.9}$$

m amount of mass (in m mol)
t thickness (in cm)
F area occupied by absorbing substance (in cm^2) hence

$$m = \frac{A \cdot F}{\varepsilon}. \tag{9.10}$$

Equation (9.10) shows that, for the determination of mass, we also need two
constants, the molar absorptivity and the area occupied by the feature under
study, but the area is much more easily determined than the thickness. The
area may be obtained by means of an ocular-micrometer scale or by means
of any area pattern of known dimension situated in the photometric field.
Also planimetry of a photomicrograph may be used for this purpose.

In natural specimens the mass, hence the absorbance, is often distributed
irregularly across the area, which results in a considerable error in the measured
absorbance if the irregular distribution is within the photoelectric beam
(Chap. 12, Sect. 4.2). If there is sufficient radiant flux, we can divide the area
to be measured into a series of partial areas (spots), each of which has more
or less uniform absorbance and obtain the mass across the whole area by
summing the masses in the spots as

$$m = m_1 + m_2 + \cdots + m_n = (A_1 \cdot F_1 + A_2 \cdot F_2 + \cdots + A_n \cdot F_n) \cdot \frac{1}{\varepsilon}. \tag{9.11}$$

(The subscripts specify the individual spots). If the spots are of equal size;
thus $F_1 = F_2 = F_3, = F$ the total amount of mass is expressed by

$$m = \frac{F}{\varepsilon} \cdot \sum_{i=1}^{n} A_i \tag{9.12}$$

A_i absorbance of spot F_i
F area for the measurement in the stage object

Equation (9.12) shows that, for the specified conditions, the amount of mass is proportional to the sum of the absorbances of material in the individual spots. The measuring values for the spots are most conveniently collected with the help of area-scanning procedures (Sect. 6.2).

9.5.4 Refractive Index and Absorption Coefficient Derived from Reflectances

In absorbing substances, such as metals and ore minerals, the refractive index (n) and quantities related to absorption cannot be measured directly but only derived from reflectances measured in two media; the absorption is most conveniently expressed by the absorption coefficient (k) (see definition of quantities in App. 2.4). In practice, the specular reflection at normal incidence is measured and the following equation is used for calculation

$$^xR = \frac{(n-N)^2 + k^2}{(n+N)^2 + k^2}$$ (9.13)

R specular reflectance at normal incidence of light (fraction)
n refractive index of reflecting material
k absorption coefficient of reflecting material
N refractive index of the medium into which the light is reflected (The prescript x of R indicates this medium, such as immersion oil, water, etc., e.g. ^{oil}R)

Note: When no medium is indicated in this way it is understood to be air, hence N is taken to be 1.

Equation (9.13) is graphically represented for air and immersion oil by Figure 9.6. It is important to state that in a given medium a certain reflectance is determined by a certain pair of values of n and k, but that the number of pairs of n and k, related to a certain reflectance is infinite. On the other hand, two reflectances measured in two different media, e.g. air and immersion oil, by their solution yield n and k unequivocally.
Transformation of Eq. (9.13) gives,

$$n = \frac{\frac{1}{2}(N^2 - 1)}{N\dfrac{1 + {}^xR}{1 - {}^xR} - \dfrac{1 + R}{1 - R}},$$ (9.14)

$$k = \sqrt{\frac{R(n+1)^2 - (n-1)^2}{1 - R}}$$ (9.15)

R reflectance in air (fraction)
^xR reflectance in the medium x (fraction)
N refractive index of the medium x (of course, all parameters must be specified by the wavelength)

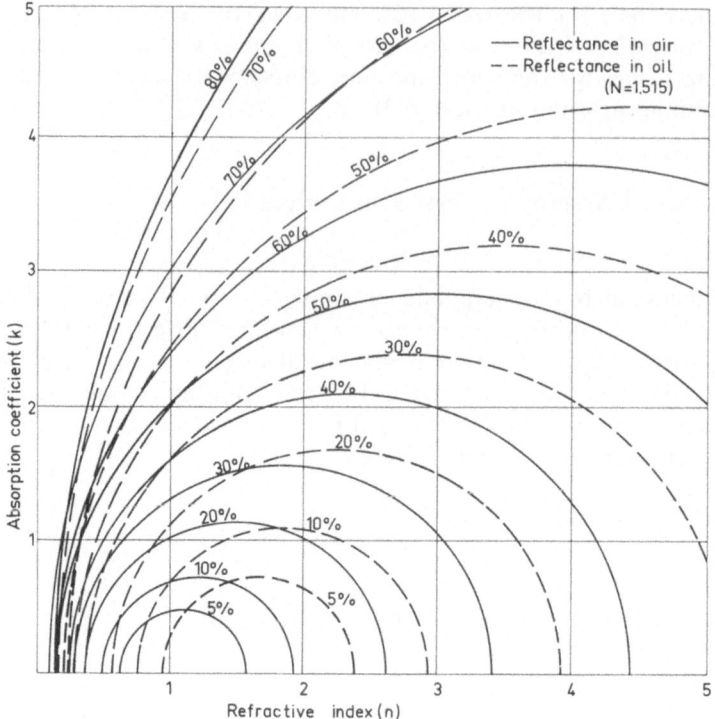

Fig. 9.6. Graphical representation of Beer's formula (isoreflectance curves). The formula is

$$^{x}R = \frac{(n-N)^2 + k^2}{(n+N)^2 + k^2} \qquad (9.13)$$

^{x}R specular reflectance of the material to be measured at normal incidence (fraction); n refractive index of the specimen; k absorption coefficient of the specimen; N refractive index of the medium into which reflection takes place. (The superscript in front of R, such as ^{oil}R, specifies the medium, if it is other than air).

The loci of reflectance are semicircles the centres of which have the coordinates $k=0$ and $n = N \cdot (1 + {}^{x}R)/(1 - {}^{x}R)$ and the radii of which are $r = 2N\sqrt{{}^{x}R}/(1 - {}^{x}R)$.

Full lines: reflectance in air $(N=1)$; *broken lines:* reflectance in oil having $N=1.515$

Notes:

1. For the graphical determination of n and k, we read off the coordinates of the point of intersection of the two appropriate reflectance curves.

2. The figure shows qualitatively that there are areas in which the curves intersect at a low angle so that the points of intersection cannot be determined accurately. It is necessary to remember this when calculating because, no matter how accurate the measuring result may be, there is an intrinsic lack of accuracy for n- and k-values, when the curves cross in these areas. (For mathematical description see Piller and v. Gehlen, 1964)

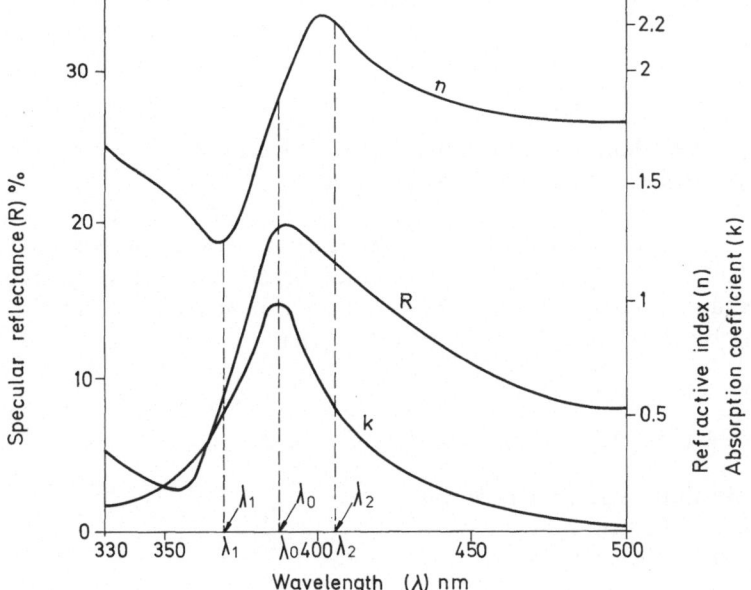

Fig. 9.7. Typical inter-relations between the spectral distribution of specular reflectance at vertical incidence, absorption coefficient and refractive index. The curves are reconstructed from those published by Anex (1966) for a crystal of auramine perchlorate for the vibration direction of minimum reflectance. (The shape of the reconstructed curves is idealised.)

The typical behaviour of such curves can be qualitatively described as follows: Near a region of maximum reflectance there is a region of maximum absorption; the peak wavelength for reflectance is somewhat longer than that for the absorption coefficient (λ_0). At the wavelength on the short side (λ_1) where the absorption coefficient has half its maximum value the refractive-index curve changes direction. Away from the immediate region of a peak of the refractive-index curve (λ_2) the trend of the refractive-index dispersion is described as normal when the refractive index decreases with increasing wavelength, and abnormal when it is the other way.

Note: In most cases, the inter-relations are not as obvious as in this figure because of the overlapping of several absorption bands

Note: If we construct the spectral curves (dispersion curves) of reflectance, refractive index, and absorption coefficient we shall find that the curves run according to certain rules as shown in Figure 9.7; but there are also spectral curves the typical interrelation of which is suppressed by the overlapping of several absorption bands.

With the help of Eq. (9.13) we can derive the refractive index of non absorbing microscopic material from the reflectance unless the refractive index is measured directly. This procedure is, for example, applied to glass particles or crystalline substances in polished sections or in mounts and to substances the refractive index of which is too high (higher than about 1.8) so that it cannot be measured with a refractometer or by using immersion methods (Koritnig, 1974). We take $k = o$ and express the reflectance as

$$^xR = \left(\frac{n-N}{n+N}\right)^2,$$

(9.16)

hence the refractive index of the reflecting material as,

$$n = \frac{N(1 + \sqrt{^x R})}{1 - \sqrt{^x R}} \tag{9.17}$$

$^x R$ reflectance in a medium with refractive index N (fraction)
(For reflectance in air it is true $n = (1 + \sqrt{R})/(1 - \sqrt{R})$)

On the other hand we can calculate the reflectance of a substance in a medium other than air if we know its reflectance in air and the refractive index of the medium in consideration. For this purpose we replace n in Eq. (9.16) by the term on the right side of Eq. (9.18), so that

$$^x R = \left(\frac{1 + \sqrt{R} - N(1 - \sqrt{R})}{1 + \sqrt{R} + N(1 - \sqrt{R})} \right)^2 \tag{9.18}$$

x   specifies the medium, e.g. immersion oil
R   reflectance in air (fraction)
(See also Jacob and Knickrehm, 1975)

Note: In practice, Eqs. (9.17) and (9.18) can be applied also to very weakly absorbing material without introducing a measurable error. 'Very weakly' means that the absorption coefficient does not exceed about 0.001.

9.6 Automation in Measuring and Data Processing

9.6.1 General Aspects

No attempt will be made in this book to review the immense number of publications on automation of photometry with the microscope, which is expanding exponentially at the present time; only the most important aspects will be mentioned. For further study the following literature is recommended: Evans (1968), Müller (1972), Wied and Bahr (1974), Galbraith et al. (1975), Rosenfeld (1975), Davis and Vastola (1976), British Standard 3527.

Automation is very useful where a large number of measuring results must be processed in a short time. It does not, of course, compensate for poor quality in the microscope image or for incorrect adjustment of the equipment. Automation does not replace all human activity in this field, but merely complements this by achieving quick handling of the data. Extensive preparation for using automation is required, and work in an air-conditioned environment at constant temperature is desirable. It must always be kept in mind that only continual use of automatic equipment can justify its expense.

The simplest case of automation is that which consists of electronic and electromechanical hardware[6] with logic circuits specially designed for the micro-

---

[6] Hardware comprises all the actual machinery, electronic circuits, etc. which carry out the various operations in computing and controlling. Software comprises all the programming and instructions that are put into the computer either beforehand or in the performing of a particular task.

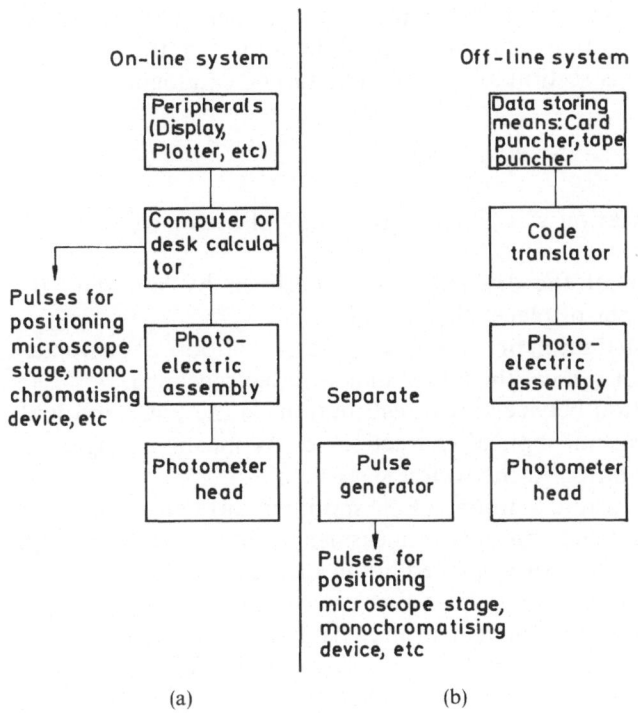

Fig. 9.8a and b. Block diagrams of essential peripherals linked to the photoelectric assembly of the microscope photometer.
  (a) System for on-line operation with a computer. (b) System for off-line operation or with data-storing means

scope photometer in consideration, this controls the movement of the microscope stage and of the monochromator and indicates the individual measuring results by printing out, displaying on a screen, or plotting.

More versatile automation is achieved by means of either a desk calculator or a large computer of standard design, linked through an electrical interface to the measuring equipment. There is no sharp distinction between a desk calculator and a computer, and in what follows we shall use the latter term for both. In general, however, the larger the computer the smaller is the amount required of special hardware for automation and the greater the amount of control that can be done by software.

If the computer is linked directly to the photoelectric assembly, this is called on-line operation (Fig. 9.8a). If the computer is separate from the photoelectric assembly, this is called off-line operation (Fig. 9.8b). (For concepts in computer technique see Sippl, 1966.) In on-line operation the computer is directly linked to motors, it sends instructions to operate these. With off-line operation it is not possible to do anything more than automatic data processing; for this some kind of data-storing device, such as a card puncher, is required.

The computer may be linked to standard peripherals (display screen, printer, coordinate plotter, etc.) that enable the measuring results to be displayed so that they can be immediately and conveniently interpreted in diagrams; this can be done in the form of histograms, frequency or size distribution functions, spectral curves, maps showing areal distribution of parameters or threshold areas. Examples for such graphs may be found plentifully in the literature listed at the beginning of this section and in manufacturers' catalogues.

9.6.2 Area Scanning

A strong impetus for automation was given by the study of biological materials in the following ways:

1. Rapid determination of the dry mass in a specimen by summing the amounts of mass in a number of places [Eq. (9.12)].

2. Detection of areal distribution or 'contiguity analysis' of optical characteristics, such as absorbance or optical thickness, in a variable specimen.

3. Detection of correlation between areal distribution on the one hand and, on the other, morphological and optical characteristics by means of logic circuitry; this is pattern recognition in the widest sense.

These procedures involve rapid scanning of the specimen through the measuring beam, the processing of a large number of measuring values, and the storage of position coordinates for each spot measured in the specimen.

The stage is moved continuously, or else stepwise in two mutually perpendicular directions in its own plane. The movement may be along a set of equidistant parallel straight lines, in a meander pattern, or in a comb pattern. Another type of movement consists of rotation about the stage axis coupled with a linear movement in the plane of the stage, which gives a spiral movement.

9.6.3 Wavelength Scanning

In very recent times the computer has become important in the handling of spectral distribution of measuring results (Piller and Prager, 1972; Van Gijzel and Schwirtlich, 1977). The reference values for each of the pre-selected wavelengths are stored; in transmitted light there is no spectral change in the reference value, but in reflected light the spectral values of the actual standard material must be inserted. Then the measuring system is operated with wavelength change for both stage objects in turn. As measurement proceeds, the measuring values for specimen and reference material at each selected wavelength are automatically fed into the computer. The input data are immediately compared by the computer with the stored reference data and the measuring results calculated and exhibited in some desired way. In the simplest case the setting of the monochromatising device is done manually and the instruction to operate the measurement is given by the operator himself. But the whole series of operations can be automated so that the computer also controls the setting of the monochromatising device and selects the wavelengths for the measuring values.

CHAPTER 10

Special Techniques in Microscope Photometry

10.1 Microscope Fluorometry

10.1.1 Introduction

Luminescence in a body is the emission of light on irradiation with a beam of shorter wavelength. When this effect ceases with the irradiation (the response time being less than 1 μs) the phenomenon is called fluorescence. When there is an after effect, it is called phosphorescence. For detailed information the following literature may be consulted: Pringsheim (1949, 1951), Passwatev (1967).

The present section is concerned with the measurement of fluorescence in microscopic objects, called microscope fluorometry. Of course, in order to obtain measuring results, the operator must be familiar with the general techniques of fluorescence microscopy. This is described in manufacturers' pamphlets and many publications; e.g. Thaer (1966), Alpern et al. (1975), Ruch and Leemann (in preparation).

In microscope fluorometry three topics are of interest:

1. Both excitation and fluorescence in the UV range. Since the two wavelengths are about 250 nm and 350 nm respectively, only a UV-light transmitting microscope can be used along with the appropriate types of lamp and photocathode.

2. Excitation in the UV range and fluorescence in the visible. In this case a UV lamp and UV optics in the illuminating system of the microscope are required, but the imaging system can be normal, while the photocathode need be sensitive only for the whole of the visible range.

3. Excitation in the short end or the middle of the visible and fluorescence in the long end. Normal optics only are needed, but again the photocathode must be sensitive over the visible range, especially at the long wavelength end.

10.1.2 Conditions of Measurement

Formerly both transmitted and reflected light were used in the measurement of fluorescence, but it has been recognised that the latter is the better technique. It is now desirable to carry out such measurements with proper control of the factors involved in the measuring value. These are the exciting light, filters, quality of the reflector, nature of the objective, and the adjustment of the diaphragms.

In carrying out measurements of fluorescence the following points have to be kept in mind:

1. The energy of fluorescent light is much less than that of normal transmitted or reflected light.

2. Fluorescence in glass and in mounting media through which the beam passes can be superimposed on the effect arising from the stage object.

3. The fluorescence effect depends not only on the specimen but also on the spectral composition of the exciting radiation (Fig. 9.4).

4. The intensity and spectral distribution of the fluorescence may change rapidly while the specimen is being illuminated (alteration effect).

5. The intensity of fluorescence is dictated independently by both the optical flux of the illuminating and the imaging system (Chap. 3, Sect. 4.3).

In view of these variables, strict standardisation and adjustment of the equipment is essential, and we may note the following requirements for good measurements. In all such relative measurements, great care must be taken to ensure that specimen and reference material are measured under exactly the same conditions.

1. Normal-incidence bright-field illumination should be used, with as large optical flux as possible; the last stage lens in the illuminating lens is, of course, the objective itself. A large optical flux is obtained by using a fluorite objective or an apochromat having a higher numerical aperture than the corresponding achromat and for the same reason, the photometric field, and thus the luminous field, should be large but should not exceed the feature to be measured in order to avoid illumination of other features causing alteration in fluorescence of those features prior to the measurement. Of course, the high numerical aperture is useful only if the aperture area is fully illuminated.

2. All equipment, including immersion media, must be selected so as to be free of any fluorescent effect.

3. Three kinds of measurements are made:

a) The integral intensity across the whole fluorescence spectrum; in this case the barrier filter must transmit and the photocathode be sensitive to the whole spectrum.

b) Only a selected spectral band; while a very sensitive photosensor must be used, there is a lower limit to the spectral bandwidth that will give a sufficiently high measuring value; the barrier filter must be 'monochromatic.'

c) Spectral distribution of fluorescence; in such measurements a continuous-spectrum monochromatising device has to be placed in the imaging light train (van Gijzel and Schwirtlich, 1977), and the photocathode must have a suitably wide range of adequate response; alternatively, the photosensor may be replaced for use with different wavelength regions. Spectral distribution of fluorescence must be specified by the spectral distribution of the exciting light, or the emission peak of a high-power spectral lamp.

As mentioned above, many specimens show an alteration in the fluorescence under illumination; this can happen within tenths of a second in some cases, and the process is irreversible. In such case the specimen must be protected from light until the measurement is made, and the measuring value must be taken immediately; it follows from this condition that all pre-adjustments have to be made without illuminating the specimen. For the rapid display of measured values an oscilloscope or oscillograph is most suitable. Electromechanical systems are now on the market that enable a quick change to be made from observation and adjustment in low-power transmitted light to fluorescence measurement in reflected light or vice versa; these are automatically controlled to fulfill the required measuring conditions (Ruch and Trapp, 1974).

10.2 Photometric Determination of Optical-Path Differences

10.2.1 General Procedures

In transmitted light the optical-path difference is defined by

$$OPD = n_1 \cdot t_1 - n_2 \cdot t_2 \tag{10.1}$$

OPD optical-path difference (in a length unit; the same as that of t, mostly µm or nm)
n refractive index
t mechanical thickness (the subscript numbers specify the media to be compared)

In practice, one medium is usually the specimen and the other the surrounding embedding material, so that $t_1 = t_2$, and

$$OPD = t \cdot (n_1 - n_2). \tag{10.2}$$

If the thickness may vary across the specimen, so will the optical-path difference.
Alternatively, and more usually, we define the optical-path difference as

$$OPD = \frac{\Delta \varphi \cdot \lambda}{2 \pi \, \text{rad}} = \frac{\Delta \varphi \cdot \lambda}{360°} \tag{10.3}$$

$\Delta \varphi$ phase difference of light waves traversing the paths in consideration (in radians or degrees)

In this equation the optical-path difference is expressed as a fraction or a multiple of a wavelength (λ).

The measurement applies only to specimens and media that are perfectly transparent; from it we can derive either n or t, provided that the other of these is already known.

The commonest measuring method is to employ calibrated compensators where an optical interference is produced; alternatively, the displacement of interference fringes may be measured (Krug et al., 1961; Beyer, 1974). The employment of the microscope photometer is a new field that has been stimulated by the recent improvement in equipment. It is specially useful in the rapid

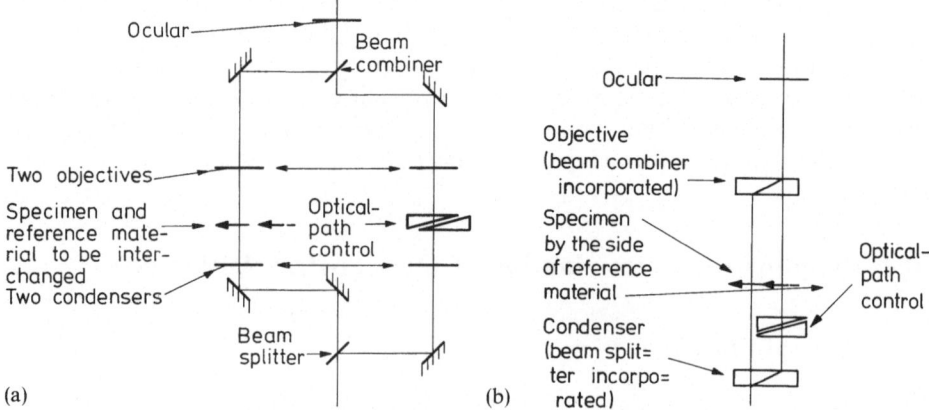

Fig. 10.1a and b. Interference systems linked to microscopes, schematic.
      (a) Double-beam interference system with large shearing (Mach-Zehnder type). (b) Double-beam interference system with small shearing (Jamin-Lebedeff type).

Notes:

1. In both (a) and (b) one of the possible arrangements of modules forming the interference device is shown. In each the whole device may be placed at a different level of the microscope so that the beam splitter comes in front of the condenser (or the stage object) and the beam combiner in front of the primary image; alternatively the beam splitter may come in front of the primary image and the beam combiner in front of a second real image. The figures show the levels of optical elements for the former arrangement.

2. In systems operating with polarised light the polariser is placed in front of the beam splitter and the analyser behind the beam combiner; in the drawings polarising devices are omitted.

3. In (a) two light trains with identical optical systems (double microscope) are required; in (b) both trains make use of the same optics.

4. There is a great variety in technical design of both kinds of interference equipment for which detailed information may be obtained from manufacturers' catalogues

determination of variations of optical path across a field since scanning procedures may be utilised (Carlson et al., 1970; Boguth, 1974).

      A double-beam interference system is required for this kind of work. The interference phenomena are observed in the image plane, while the image is repeated in the plane of the photometer diaphragm. Model systems of interference equipment are shown in Figure 10.1.

      It is best to restrict the use of such methods to cases where the amount of the difference in optical path between the specimen and the reference medium is less than $\lambda/2$; otherwise ambiguities arise in the interpretation of intensities.

10.2.2 Correlation of Photometric Values and Optical-Path Difference

The following steps are carried out in making the measurement:

      1. With monochromatic light the optical path of the auxiliary beam is regulated so as to produce maximum brightness in the reference material, e.g. the embedding medium, which gives the adjustment for perfectly constructive interference; thus the waves traversing the two optical paths are in phase. A device with continuously variable path length is used for this purpose.

2. The measuring value (photometric value) for the interference effect of the reference medium is then established.

3. After this, the optical path of the auxiliary beam is regulated so as to produce minimum brightness (maximum extinction) in the reference material, which gives the adjustment for perfectly destructive interferences, thus the measuring value for background light (glare) and other effects causing incremental values. If this value is of considerable amount—for example larger than one hundredth of the former value—it causes an 'incremental error,' and must be subtracted from all measuring values for the reference material and the specimen (Chap. 12, Sect. 3.1 and App. 5).

4. Leaving the adjustment untouched, the reference medium is replaced by the specimen, and its measuring value in interference is established.

5. The measuring result is obtained from the relation

$$\frac{G_{sp}}{G_{ref\,max}} = \cos^2\left(\pm\frac{\Delta\varphi}{2}\right) = \cos^2\left(\pm\frac{OPD\cdot\pi\,\text{rad}}{\lambda}\right) = \cos^2\left(\pm\frac{OPD\cdot 180°}{\lambda}\right) \quad (10.4)$$

G measuring value eventually corrected in respect to an increment (numeral)

sp   specimen

ref max   reference material with interference system adjusted so that maximum brightness is achieved

$\Delta\varphi$ phase difference of light waves traversing the specimen or the reference material (in radians or degrees)

OPD optical-path difference expressed by Eq. (10.3)

λ wavelength

[For derivation of Eq. (10.4) see App. 7].

Notes:

1. If it is necessary to know whether the specimen has a smaller or a greater optical path than the reference material, and so a smaller or greater refractive index, the Becke-line test should be applied (App. 9); alternatively the specimen should be viewed in 'phase contrast' (with phase contrast equipment in the microscope), in which case the higher refracting medium appears darker than the lower refracting one (Barer, 1956).

2. If it is necessary to know whether the optical-path difference to be measured is greater than the permitted amount ($\pm \lambda/2$) then, prior to the photometric procedure, it must be estimated using classical methods, e.g. measurement of the displacement of interference fringes or with the help of a calibrated compensator. Instructions for this method may be obtained from textbooks, e.g. Beyer (1974).

3. Care must be taken to see that the thicknesses and refractive indices of the media above and below the specimen and the reference material in the photometric beam are the same; cover slip, embedding medium, specimen slide.

10.2.3 Correlation of Optical-Path Difference and Refraction Parameters

The parameters that can be derived from optical-path difference are listed in Table 10.1. In the following the derivation is described.

Table 10.1. Equations for the derivation of material parameters from optical-path differences measured in transmitted light

Specification of measured optical-path difference	Derived parameters	For the derivation of parameters listed in column 2, the following parameters must be known	Defining equation	Conditions for the derivation
Optical-path difference in the two media 1 and 2; (OPD)	Refractive index of medium 2; (n_2)	Refractive index of medium 1; (n_1), thickness of media 1 and 2; $(t_1 = t_2 = t)$	$n_2 = n_1 + \dfrac{OPD}{t}$	$n_1 < n_2$
			$n_2 = n_1 - \dfrac{OPD}{t}$	$n_1 > n_2$
	Refractive index of medium 1; (n_1)	Refractive index of medium 2; (n_2), thickness of media 1 and 2; $(t_1 = t_2 = t)$	$n_1 = n_2 + \dfrac{OPD}{t}$	$n_1 > n_2$
			$n_1 = n_2 - \dfrac{OPD}{t}$	$n_1 < n_2$
	Thickness of media 1 and 2; $(t_1 = t_2 = t)$	Refractive indexes of media 1; (n_1) and 2; (n_2)	$t = \dfrac{OPD}{n_2 - n_1}$	$n_1 < n_2$
			$t = \dfrac{OPD}{n_1 - n_2}$	$n_1 > n_2$
Optical-path difference in the media 1 and 2; $OPD_{(1/2)}$ and in the media 1 and 3; $OPD_{1/3}$	Refractive index of medium 1; (n_1)	Refractive indexes of media 2; (n_2) and 3; (n_3)	$n_1 = \dfrac{n_2 \cdot OPD_{1/3} - n_3 \cdot OPD_{1/2}}{OPD_{1/3} - OPD_{1/2}}$	$n_1 < n_2$ $n_1 < n_3$
			$n_1 = \dfrac{n_3 \cdot OPD_{1/2} - n_2 \cdot OPD_{1/3}}{OPD_{1/2} - OPD_{1/3}}$	$n_1 > n_2$ $n_1 > n_3$
			$n_1 = \dfrac{n_2 \cdot OPD_{1/3} - n_3 \cdot OPD_{1/2}}{OPD_{1/2} + OPD_{1/3}}$	$n_1 < n_2$ $n_1 > n_3$ or $n_1 > n_2$ $n_1 < n_3$

Thickness of media 1, 2 and 3; ($t = t_1 = t_2 = t_3$)	Refractive indexes of media 2; (n_2) and 3; (n_3)	$t = \dfrac{OPD_{1/3} - OPD_{1/2}}{n_3 - n_2}$	$n_1 < n_2$ $n_1 > n_3$
		$t = \dfrac{OPD_{1/3} - OPD_{1/2}}{n_2 - n_3}$	$n_1 > n_2$ $n_1 > n_3$
		$t = \dfrac{OPD_{1/3} + OPD_{1/2}}{n_2 - n_3}$	$n_1 < n_2$ $n_1 < n_3$
		$t = \dfrac{OPD_{1/3} + OPD_{1/2}}{n_3 - n_2}$	$n_1 > n_2$ $n_1 < n_3$
Optical-path difference in a gel or solution and the pure solvent; (OPD)	Amount of mass of solved substance; (m) Area occupied by the gel or solution; (F) Specific refraction increment of gel or solution; (α)	$m = \dfrac{OPD \cdot F}{100 \cdot \alpha}$	Units: OPD in cm^2 F in cm^2 α in $cm^3 \cdot g^{-1}$ m in g c in $g \cdot cm^{-3}$
Concentration of solved substance; (c)	Thickness (t) and specific refraction increment of gel or solution; (α)	$c = \dfrac{OPD}{t \cdot 100\alpha}$	

Notes:

1. The lengths (thicknesses, wavelengths) and optical-path differences correlated with lengths must be given in the same unit.

2. For all equations the media, to which the path difference is related, are taken to have equal (mechanical) thickness (t). This condition is automatically fulfilled when the medium to be investigated is completely surrounded by the reference medium, and the surfaces of the media through which the photometric beam is traversing, such as the surfaces of parallel plates are normal to the direction of the beam axis.

3. For photometric determination of optical-path differences the material to be compared must be clear and perfectly transmitting.

Refractive Index and Thickness. Measurement of the optical-path difference is a means of deriving one of the three quantities on the right side of Eq. (10.2). Hence we must know two of the quantities in order to calculate the third one. The measurement of two optical-path differences for the same specimen in two surrounding media of different known refractive indices allows us to determine both the thickness and the refractive index of the specimen. The defining equations are given in Table 10.1.

Concentration and Amount of Refracting Material in a Gel or Solution. In a gel or solution the refractive index is often related to the concentration and so the latter may be found by determining the former; this applies especially to living cells, chromosomes and tissues in biology. The basic correlation can be set out as follows:

Proportionality between refractive index and concentration exists usually only for dilute solutions; the factor is called the specific refraction increment and is defined by

$$\alpha = \frac{n_g - n_s}{100 \cdot c} \tag{10.5}$$

α specific refraction increment (in $cm^3 \cdot g^{-1}$); the term $(\alpha \cdot c_0)$ means the change in refractive index with 1% change in concentration (c_0).
n refractive index
g gel or solution
s pure solvent
c concentration of dissolved material (in $g \cdot cm^{-3}$)

Notes:
1. Of course, α and n are functions of wavelength.

2. In Eq. (10.5) the mass of dissolved material is expressed in the unit of g, not in that of mol, as in Eq. (9.9); this fact results from the definition of α.

3. For material in biological specimens it is found that the optical-path difference usually does not exceed $\lambda/2$.

4. Values for α are given in the literature, e.g. Davies (1958).

We rearrange Eq. (10.5), replace the terms n_1 and n_2 in Eq. (10.2) by n_g or n_s respectively and express $(n_g - n_s)$ by (OPD/t) in order to obtain the concentration,

$$c = \frac{OPD}{t \cdot 100\alpha} \tag{10.6}$$

OPD optical-path difference in the gel or solution and the pure solvent (in cm)
t thickness of the gel or solution (in cm)

The amount of mass of dissolved material with the unit taken as the gram is obtained by multiplying the concentration by the volume of the gel or solution so that

$$m = \frac{OPD \cdot F}{100\,\alpha} \tag{10.7}$$

m mass of dissolved material (in g)

F area in the stage-object plane occupied by the gel or solution (in cm^2)

10.3 Diffuse Reflection from Surfaces

Diffuse reflection is due to scattering, diffraction, refraction and absorption of light by an accumulation of small particles. Particles at the surface and those in the interior of the body contribute to diffuse reflection. A fraction of the incident light may traverse the accumulation, if this is thin enough. When no light is transmitted through the substance the latter is taken to be infinitely thick and its capacity for reflection is called reflectivity (App. 2.4). Furthermore, the reflected light may have a specularly reflected component and this may have a preferred direction according to the orientation of specularly reflecting surface elements (gloss). Hence diffuse reflection is a very complex phenomenon; its mathematical description is rather complicated (Kortüm, 1969) and beyond the scope of this book.

In microscope photometry it is important to state that, for a given stage object, the amount of diffusely reflected light depends on: (1) the geometry of the incident light beam, (2) the angle of the light cone that the objective is able to accept (Fig. 10.2), (3) the reflection properties of the substrate on which the stage object is placed, unless the stage object can be considered to be infinitely thick.

Note: Perfect diffuse reflection occurs only with isometric (e.g. globular) particles and with isotropic (optically) ones; the particles must have a diameter somewhat larger than one wavelength of visible light and must be uniformly distributed.

In view of the variables (1) to (3) strict standardisation and adjustment of the equipment is essential, and we may note the following requirements

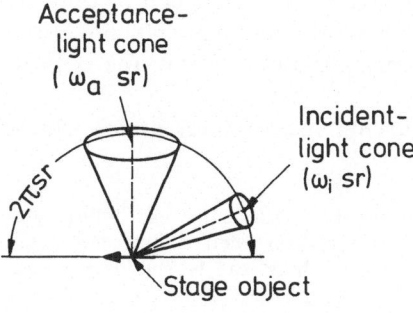

Acceptance-light cone (ω_a sr)

2π sr

Incident-light cone (ω_i sr)

Stage object

Fig. 10.2. Geometry of illumination and light emission on diffuse reflection. In diffuse reflection the incident light consists of a cone of specified angle (ω_i sr). When this is reflected uniformly over half the complete solid angle (2π sr) the shape of the diffuse-reflectance indicatrix is a hemisphere. Thus for measurement of the total amount of diffusely reflected light we would require an optical system capable of accepting light over a hemisphere (integrating sphere). In microscopy, however, this cannot be done, as the acceptance cone is limited by the potential numerical aperture of the objective; ($\omega_a = \pi \cdot NA_{Obj}^2$). This must be kept constant in comparison measurements and should be specified for reproducible measurements at the reference material as well as at the specimen to be studied.

1. Dark-field illumination should be used in order to suppress the effect of specular reflection at glass surfaces in the photometric beam (surfaces of lenses, cover glass, etc.). Of course, only measurements in air are useful. The illumination may be multilateral (Fig. 2.1 k, l) or unilateral (Fig. 2.1 m, n).

2. The geometry of the incident beam must be carefully adjusted according to the recommendations of standardising committees, or else specified (Stenius, 1955; German Standard DIN 5036). This specially concerns the obliquity of illumination, the aperture angle of the illuminating and imaging beam, and the relation between the diameter of the luminous field and the photometric field in the stage object. As a rule, the larger the diameter of the luminous field for a given diameter of the photometric field the more light is diffusely reflected at spots outside the photometric beam and superimposed on the light reflected at spots lying inside the photometric beam.

3. Stage objects that are translucent[7] must be placed on a black substrate.

10.4 Techniques for Measuring Micro-Autoradiographs

The features of micro-autoradiographs are accumulations of silver grains that arise after the development of a photographic emulsion that has been exposed to radiation from radioactive elements used for labelling certain biological materials. We can measure either with transmitted or with reflected light (Rogers, 1967; Przybylski, 1970; Dörmer, 1973). For the former the stripping-film technique has to be employed, whereby the emulsion is removed from the specimen; this has the disadvantage that it is not possible to correlate immediately details of the radiograph with details on the specimen. This difficulty does not occur with reflected-light measurement, and so we shall discuss only this here.

First we establish the relation between the measuring value and the number of silver grains actually counted on several spots of a reference substance; the calibration is expressed in an equation, plotted graphically (Fig. 10.3), or else fed into a computer. Of course, various causes such as silver grains arising from environmental radioactivity, impurities in the gelatine and structures in the underlying specimen, give rise to disturbances in the light beam that is being measured. Brightness in the background causes an incremental measuring value (App. 5). This is obtained by making several measurements on areas free from silver grains and this will contain any glare effect from the optical system.

The measuring value is affected by the varying size of silver grains which complicate the relation.

---

[7] The term translucent is applied to materials, such as paper sheets, that admit and diffuse light so that objects beyond cannot be clearly distinguished through them. Material that admits no light at all, hence appears perfectly dark if illuminated from the back, can be considered to be infinitely thick from this point of view.

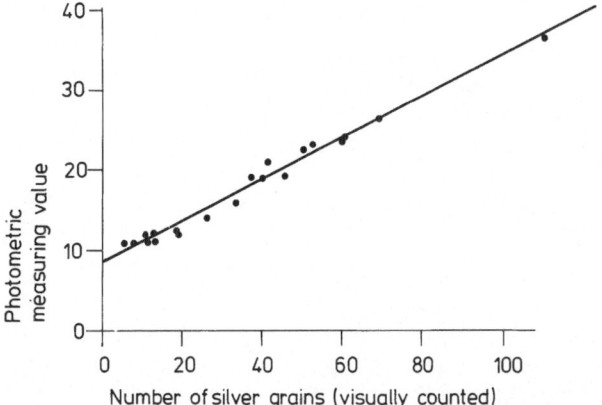

Fig. 10.3. Photometric values for accumulated silver grains in an autoradiograph film obtained on reflection with bright-field illumination (after Dörmer, 1973). The figure shows an ideal linear relationship between the measuring value and number of grains. This occurs only when the grains are uniformly dispersed and not too dense.

Notes:
1. The photómetric beam contains both diffusely and specularly reflected components.
2. The point at which the graph intersects the ordinate indicates the measuring value for brightness in the image background. Corrections according to the description in Appendix 5 may be applied

A serious source of error occurs when the grains are not evenly dispersed due to local accumulation, or when their density in the measuring beam is too large; in such cases the upper grains mask the lower and also reduce the background value component, with the result that the relation between the measuring value and the number of grains is not linear (Fig. 10.4). The conditions that are necessary for a linear relation between number of grains and measuring value are uniform size of grain, uniform dispersion and a not-excessive grain density. The linear relation is simply

$$N_{sp} = \frac{G_{sp}}{G_{ref}} \cdot N_{ref} \tag{10.8}$$

N number of grains
G measuring value
sp   specimen
ref   reference material

Where the brightness in the background is not negligible, we make the usual correction, according to Eq. (12.1).

When the relation is more complex, we have to use a correction diagram that is established experimentally (Fig. 10.4).

We have decided that reflected-light measurement should be used. With dark-field illumination there are disadvantages because the depth of light penetration is smaller than with bright field, so that we do not get such a good

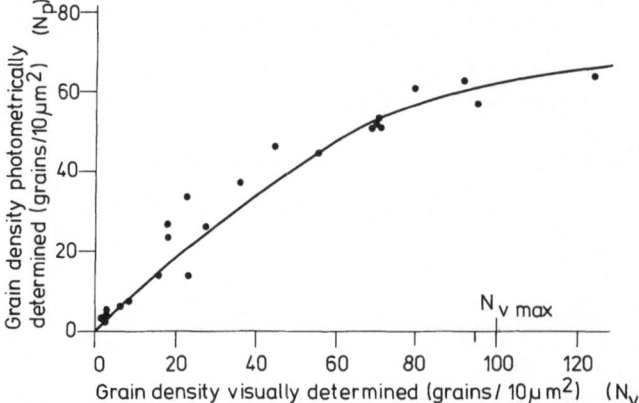

Fig. 10.4. Densities of silver grains in an autoradiograph film determined photometrically on reflection with bright-field illumination plotted against visually determined densities (after Dörmer, 1973). The graph corresponds to the following empirically determined exponential function

$$N_p = N_{v\,max}\left(1 - e^{-\frac{N_v}{N_{v\,max}}}\right) \qquad\qquad (10.9)$$

N_p photometrically determined grain density (number of grains per 10 μm²); N_v visually determined (by counting) grain density (number of grains per 10 μm²); $N_{v\,max}$ maximum value of N_v up to which determination with the photometer is allowed. Values larger than this lie in the 'range of photometric saturation'.

Note: If $N_v = N_{v\,max}$ then $N_p = N_{v\,max}(1 - 1/e)$. For the present test, $N_{v\,max}$ is ca. 100 grains per 10 μm², hence N_p is ca. 60 grains per 10 μm²

average. Further, the low level of the measuring value due to the relatively small yield of reflected light with dark-field illumination prevents the reduction of the diameter of the photometer diaphragm, hence also of the luminous-field diaphragm, nor can the latter be imaged as well as with bright-field illumination. With bright-field illumination the higher level of the measuring value allows us to reduce the diameter of the diaphragms, which reduces both disturbance and background brightness.

The only disadvantage of using bright field is the glare effect from the various surfaces in optics and specimens. But we can greatly reduce these by using an oil-immersion objective, which can be low- or medium-power, as well as high-power. Another great advantage of oil-immersion objectives in work of this kind arises from their insensitivity to the optical thickness of the gelatine layer and of the cover glass, where these are used. Reflections from the base of the slide on which the specimen is mounted can be suppressed by placing this on a layer of oil resting on black glass having a thickness of about 2 mm and a refractive index of 1.52 so that it forms a homogeneous optical body with the slide and the oil; and thus the light is absorbed in the black glass.

The light need not be monochromatic, and the use of white light is very beneficial since it raises the level of the measuring value. However, if the background light shows any coloration, then it is helpful to use a complementary-

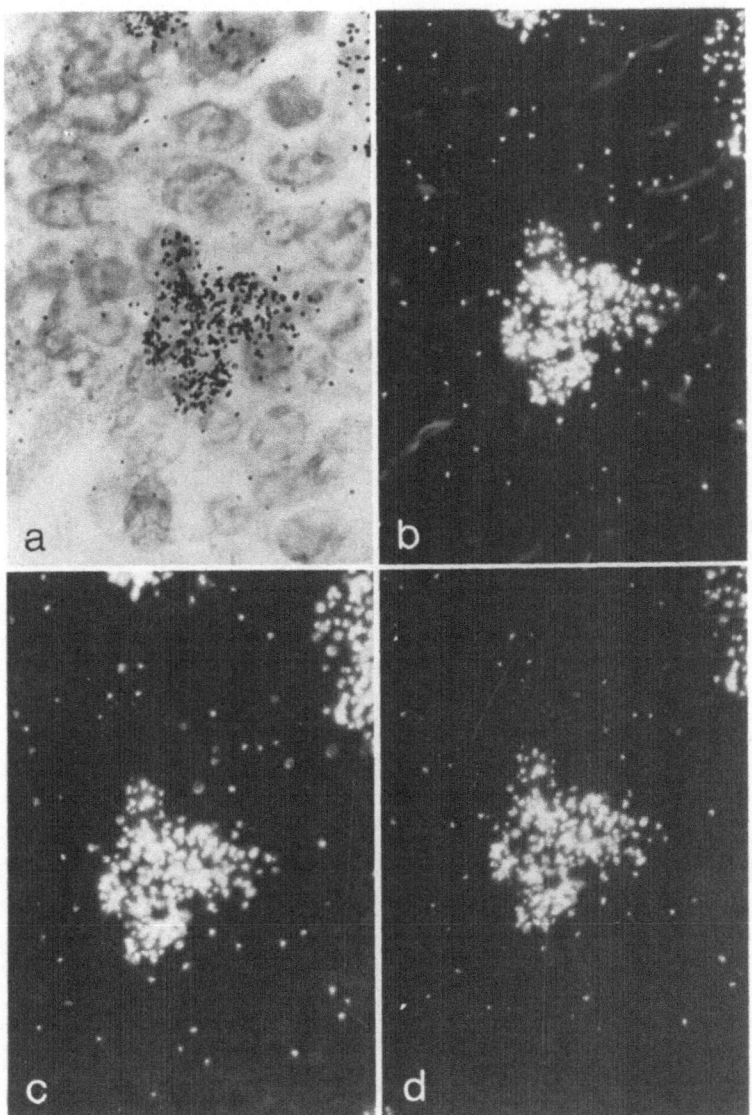

Fig. 10.5a–d. Autoradiographs from a thin section of stained liver (magnification scale approx. 500:1). (Photographs taken by the author with an objective Epiplan 40/0.65.)

(a) Transmitted-light bright field.

Note: The image of the silver-grain accumulation is significantly disturbed by underlying features.

(b) Normal reflected-light bright field.

Note: Faint features are seen in the background.

(c) Normal reflected-light bright field with black glass substrate. Same as (b) but with reduced reflections at the lower side of the specimen slide because this was placed on an absorbing substrate.

Note: The faint features in the background have largely disappeared.

(d) Same as (b) but with crossed polars.

Note: The background appears still more uniform and darker than in c colour filter of a broad-band type so as to suppress some of the background light.

An excellent way of suppressing background light is to use polarised light with the analyser inserted and crossed. The light that we wish to measure is depolarised by scattering from the rounded silver grains and so traverses the analyser; both glare and background light remain mostly polarised and so are cut out by the crossed analyser (Fig. 10.5). But, of course, the enormous loss of light so caused can only be tolerated if a lamp of extremely high radiance is used, such as a superpressure xenon lamp.

The focusing of the objective must be done with great care and checked because it is essential that all the silver grains reflecting should lie within the focal depth; this involves the use of objectives of as low power as possible and requires the magnification scale of the image in the plane of the photometer diaphragm to be as small as possible.

10.5 Micro-Densitometry

The term micro-densitometry is applied to the measurement of blackening (or density) in minute areas on developed photographs by transmitted white light. These may be ordinary photographs, photomicrographs, electron photomicrographs, X-ray diffraction photographs, spectral-line photographs, auto-radiographs, etc. Blackening is expressed formally in mathematics in the same way as absorbance [Eq. (9.5)]; we replace the term absorbance (A) by the term blackening (D_n) and the term transmittance (T) by the ratio of the measuring value from a clear, perfectly transmitting uncoated spot on the substrate of the photographic emulsion (G_{ref}) to that from the blackened area (G_n) and obtain,

$$D_n \equiv \lg \left(\frac{G_{ref}}{G_n} \right). \tag{10.10}$$

Note: The ratio G_{ref}/G_n is called opacity, while the inverse ratio is transparency.
Also, the exposure is defined as a logarithmic term, that is

$$\text{Exposure} \equiv \lg (E \cdot t) \tag{10.11}$$

E irradiance (in $W \cdot m^{-2}$) or illuminance (in lx) (for definition of terms see App. 2.5)
t time of exposure (in s)
$(E \cdot t)$ quantity of irradiation or of illumination

Hence there is a linear relationship between blackening and exposure (Fig. 10.6). This figure shows a typical blackening diagram, and it can be seen that there are parts of the curve in this diagram, in which the blackening is not proportional to the exposure: these should not be used in micro-densitometry.

Standard properties of the photoemulsion are supplied by the manufacturer, but for the most careful work these should be checked by the user for the

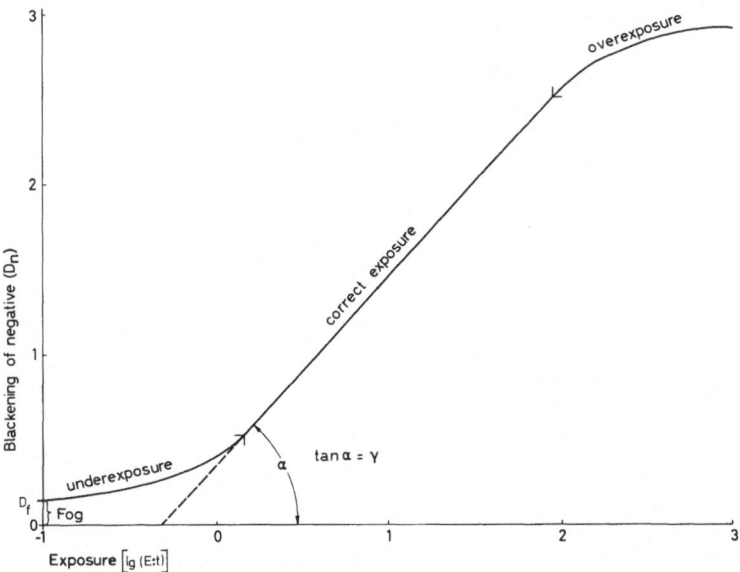

Fig. 10.6. Typical blackening diagram of a photographic emulsion. The diagram shows the blackening plotted against the exposure.

A distinction must be made between three sections of the curve; underexposed, correctly exposed (in the blackening range of about 0,5 to 2,5), overexposed. The slope of the straight section indicates the 'gradation' (γ-value) of the photoemulsion; it is expressed by

$$\gamma = \tan \alpha = \frac{\varDelta D_n}{\varDelta[\lg(E \cdot t)]} \tag{10.12}$$

γ gradation; $\varDelta D_n$ difference in blackening in the straight section of the curve; $\varDelta[\lg(E \cdot t)]$ corresponding difference in exposure.

Notes:

1. A gradation with the value 1 is considered as normal, material having a gradation less than one is considered as 'soft,' that having a gradation larger than one as 'hard.'

2. The gradation can be influenced by the development process.

3. The blackening curve never goes down to zero, but has a certain minimum value, called fog. This is due to loss of light by absorption and scattering at inhomogeneities in the non-exposed but developed emulsion. Consequently the actual blackening is expressed by

$$D_n = D_{np} + D_f = \gamma \cdot \lg(E \cdot t) + D_f \tag{10.13}$$

D_n actual (total) blackening; D_{np} pure blackening; D_f blackening due to fog (This is measured at a developed but unexposed area in the photoemulsion); $(E \cdot t)$ pure exposure; γ gradation of the photoemulsion.

For the determination of pure blackening the fog must be substracted from the total blackening

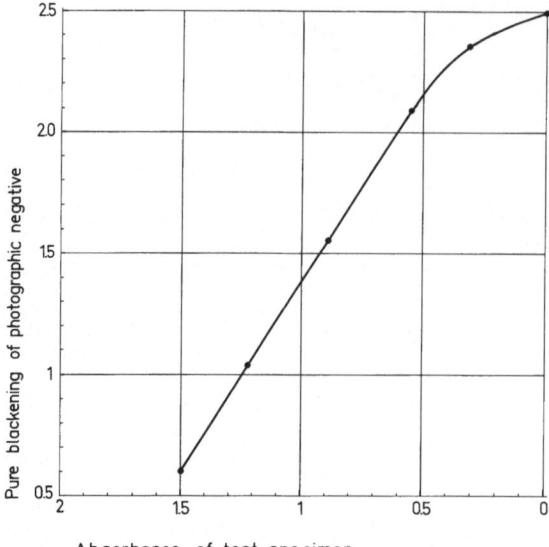

Fig. 10.7. Experimental determination of the relationship between blackening of photographic emulsion and absorbance of the photographed stage object.

A camera is mounted on the microscope and exposed to an empty stage-object field at specified conditions of adjustment of equipment. A series of photographs is taken at constant exposure time with the beam intensity reduced in known amounts by means of stepped neutral-density filters. The exposure time must be selected so that the blackening is in the limits of about 0.5 and 2.5, within which the blackening curve (Fig. 10.6) is straight. The correct time may be established by experiment; repeating the procedure described below several times for different exposures.

The experiment is started with a perfect transparent glass plate placed in the light train at a level that is suited for the insertion of filters. The experiment is continued after replacing the glass plate by neutral-density filters of known transmittance, and thus of known absorbance.

The photographic emulsion is developed under standard conditions and then placed on the microscope stage to form the stage object. The *pure* blackening is then determined for each exposure and plotted against absorbance of the neutral-density filter to which it is related. The blackening resulting from the exposure with the perfectly transmitting glass plate is that corresponding to zero absorbance. Of course, when *transmittance* of the filters is used as reference, the abscissae values are not linear.

Notes:
1. The steps in absorbance of the neutral filters need not be equal, but the range of absorbances of the filters should be the same as the range of absorbances of the actual specimen to be photographed.

In the example, the steps are from zero to 0.3, 0.6, 0.9, 1.2 and 1.5.

2. When an exposed photoemulsion is being measured it must be compared against the *uncoated* substrate as reference.

3. For the determination of pure blackening the fog must first be measured. This is done by comparing a piece of the photographic negative developed but unexposed against a piece of uncoated (clear and perfectly transmitting) substrate, e.g. film. Pure blackening results from the subtraction of fog from the total blackening according to Eq. (10.13).

4. The neutral-density filters can be replaced by stage objects of known absorbance, and photomicrographs of these are used for the test

individual batch of photographic material; the spectral composition of the light used in making the photographs and the developing conditions are the two chief causes of variation (Mees, 1959). The best way of checking is the construction of an individual graph for the actual photographic material and operating conditions (Fig. 10.7) showing the relationship between blackening in the negative and absorbance in the photographed stage object. With such a graph the blackening in the photomicrographic negative can immediately be converted to the corresponding absorbance in the photographed stage object.

Notes:

1. When the photoemulsion has a gamma value [Eq. (10.12)] greater than unity, the internal sensitivity of the measuring procedure is increased. From the diagram of blackening calibration the blackening at any point of the image can be ascribed to the corresponding part of the actual specimen as seen in an enlarged positive.

2. Micro-densitometry is often applied to the photometric study of specimens that are too small to be isolated by the photometer diaphragm. Among such are genes in chromosomes (van der Ploeg et al., 1974) but it applies to any kind of specimen which supplies a radiant flux too small to be measured. A photograph of the specimen is taken with white or monochromatic light or with an electron beam. The strongly enlarged negative image is then measured in the way that has been described.

3. Although the name microscope densitometry would fit better into the terminology of this book than the name micro-densitometry, we keep the latter name for historical reasons.

10.6 Determination of Basic Optical Constants with Interference-Layer Techniques

In Chapter 9, Section 5.4 the method of specular reflectance measurement at vertical incidence in two media is described for the determination of the two basic optical constants; refractive index and absorption coefficient. But, as shown in that section, there are areas on the isoreflectance chart in which good accuracy in determination is impossible. Of course, by changing the refractive index of the immersion medium we change these areas of poor accuracy; in this way we can often overcome the difficulty.

When a large homogeneous surface is available for measurements we can use the interference-layer method which is free (with a few exceptions) from the errors just mentioned, so far as the refractive index and absorption coefficient of metals or alloys are concerned. This method can be used also for enhancing the visual contrast between constituents in a polished section (Pepperhoff and Ettwig, 1970).

Instead of immersion in a liquid, a layer of highly refracting isotropic material is deposited from vapour on the surface of the specimen at a pressure of less than 10^{-4} torr; cubic ZnS ($N=2.39$ at 546 nm) or ZnSe ($N=2.65$ at 546 nm) have proved to be very suitable substances.

The reflectance concerned is called 'interference reflectance'; it results from the combination of three effects: reflection from the upper surface of the deposited layer, reflection from the surface of the specimen (specimen-layer interface) and interference between the light beams after reflection. For given reflectances of the layer and the specimen the interference reflectance varies with the optical-path difference of the interfering light waves, and thus with the optical thickness of the layer and with the wavelength. The interference reflectance used is the minimum within the variation range. It is called 'minimum-interference reflectance,' the layer can be considered as being an anti-reflection layer for the metal.

The condition for minimum-interference reflectance is,

$$OPD_{min} = (k + \tfrac{1}{2})\lambda \tag{10.14}$$

OPD optical-path difference of interfering light waves (in the unit of wavelength)
min   minimum
k integer including zero
λ wavelength (in nm)

The condition for the corresponding optical thickness of the layer which is traversed twice by the beam reflected at the specimen surface, is,

$$\frac{OPD_{min}}{2} = N \cdot t \tag{10.15}$$

N refractive index of the layer
t mechanical thickness of the layer

In consequence, minimum-interference reflectance occurs when the optical thickness of the layer is $\lambda/4$ or an odd multiple of $\lambda/4$.

Note: For an anti-reflection layer on a glass surface the value of k is kept zero.

The thickness of the layer can be pre-adjusted and the wavelength estimated during the deposition process. For this the specimen surface is illuminated with white light during vaporisation; the reflected light from the surface is observed through a continuous-spectrum interference filter. For a given thickness of layer there will be a particular wavelength, or for a given wavelength there will be a particular thickness of layer, for which extinction due to interference has reached its maximum. At the position corresponding to such a thickness a black band will be observed when the filter is transmitting the wavelength so defined. Detailed instruction about the systematic control of optical thickness of layers is given by the manufacturers of vaporisation equipment.

It may be desirable to continue the deposition in order to obtain a higher-order effect for the same wavelength so that a black band is seen for this wavelength and its harmonics. For observation at several places on the same specimen, a uniform thickness of deposited layer must be achieved. Alternatively, the layer may be deposited in a wedge of low angle, in which case measurements can be made at more than one wavelength.

Measuring Procedure. The measurement first requires knowledge of the wavelength for which minimum-interference reflectance occurs. This wavelength may

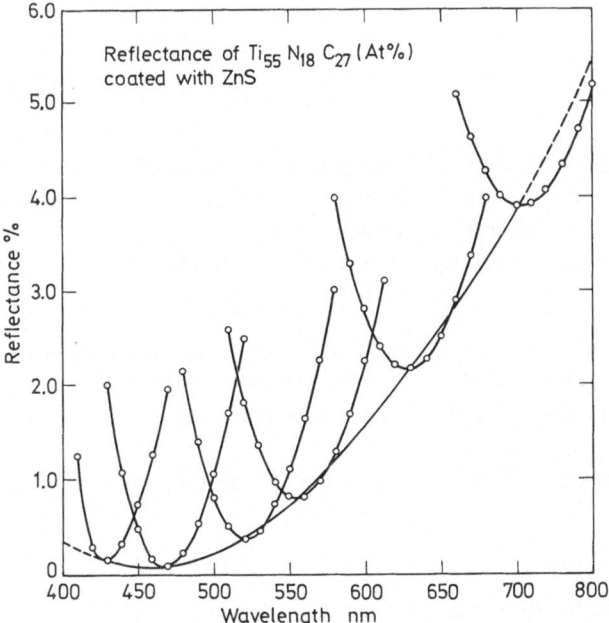

Fig. 10.8. Minimum spectral reflectances of a ZnS-interference layer on a polished surface of $Ti_{55}N_{18}C_{27(Atom\ \%)}$ (after Knosp, 1970). We have a homogeneous specimen (substrate) covered by an interference layer of varying thickness. At the first spot selected we measure the spectral reflectance arising from the interference of the beam from the substrate and that from the upper face of the layer. We note the wavelength of the minimum reflectance and the wavelength of this. This procedure is repeated for several spots chosen to correspond to different thicknesses by observation of the interference-colour pattern. We plot the measuring results as in the figure. Each parabola is the spectral-reflectance curve for a constant spot. The lowest point on the curve gives the values for minimum-interference reflectance and the wavelength at which it occurs. The spectral curve of minimum-interference reflectance is the connecting line of the lowest points from different spots.

Note: The value of the minimum-interference reflectance and of the wavelength at which this occurs is independent of the order of interference spectrum used in the measurement. However, sharper minima occur for rather higher orders of interference

be predicted according to Eqs. (10.14) or (10.15) when the optical thickness of the layer at the spot of measurement is known, or else established experimentally. The specimen is observed with the microscope during continuous change of wavelength setting. The setting is fixed at the moment when the specimen shows maximum darkness. The visual observation may be replaced by the indication of photometric values from the surface, so that maximum darkness is indicated by the minimum measuring value (Fig. 10.8). In order to measure minimum-interference reflectance for other wavelengths other spots for the measurement must be selected on the wedge.

The actual measurement consists in the comparison of the measuring value for minimum-interference reflectance with that for the reference material according to

$$R_{\text{sp min}} = \frac{G_{\text{sp min}}}{G_{\text{ref}}} \cdot R_{\text{ref}} \tag{10.16}$$

$R_{\text{sp min}}$ minimum-interference reflectance of the specimen (fraction)
$G_{\text{sp min}}$ measuring value for minimum-interference reflectance
R_{ref} reflectance of reference material in air (fraction)
G_{ref} measuring value for the reflectance of reference material

Notes:
1. For the measurement of minimum-interference reflectance, of course, the measuring equipment must be adjusted in the same way as for the measurement of normal specular reflectance.

2. It is beneficial to use a reference material that has a low reflectance such as neutral glass (Chap. 9, Sect. 3.2).

The basic optical constants are finally calculated with the help of following equations,

$$n = \frac{\frac{1}{2}(N_1^2 - 1)}{N_1 \cdot \dfrac{(1 + R_{\text{min}}) \cdot (1 + \rho_1) \pm 4\sqrt{\rho_1 \cdot R_{\text{min}}}}{(1 - R_{\text{min}}) \cdot (1 - \rho_1)} - \dfrac{1 + R}{1 - R}}, \tag{10.17}$$

$$k = \sqrt{\frac{R \cdot (n + 1)^2 - (n - 1)^2}{1 - R}} \tag{9.15}$$

n refractive index of specimen
k absorption coefficient of specimen
N_1 refractive index of interference layer
R_{min} minimum-interference reflectance of the specimen (fraction)
R reflectance of the specimen in air (fraction)
ρ_1 reflectance of the interference layer (fraction) calculated according to

$$\rho_1 = \left(\frac{N_1 - 1}{N_1 + 1} \right)^2 \tag{10.18}$$

For further information on the derivation of Eq. (10.17) and theory of interference-layer techniques see Appendix 8.

CHAPTER 11

Special Adjustment for Photometry

11.1 Focusing of Objective

Incorrect focusing of the objective does not only result in a bad quality of the image but also in a wrong photometric measuring value. Tests about the relationship between incorrect focusing positions and variations of measuring values at photometric measurements with transmitted and reflected light were carried out by the author of this book, and these gave the following qualitative results (Fig. 11.1):

1. The effect of defocusing on the measuring value depends on the design of the objective. With those objectives used in the tests it was found that to a certain extent the relative difference of measuring values is proportional to the distance between the stage object and the plane on which the objective is actually focused. The relative difference here means between the values measured in focus and out of focus related to those in focus.

2. This relative difference in measuring value does not significantly depend on the transmittance or reflectance of the stage object.

3. When the stage object is out of focus, either up or down, the measuring value may increase or decrease; it follows that the correct position cannot be obtained simply by adjusting so as to maximise the measuring value.

4. For a given amount of defocusing the relative difference of measuring value increases with increase in either the magnification or the numerical aperture of the objective. Thus, with high-power objectives the slightest defocusing causes a noticeable change in the measuring value.

5. With transmitted light the effect of defocusing is less than with reflected light.

The results of the tests are shown in Figure 11.1 a and b. From these results we can deduce the best procedure for minimising the influence of defocusing on the measuring result. Since the nature of the stage object is not noticeably involved, we must simply preserve exactly the conditions of measurement between specimen and reference material, this is what is important and not the exact correctness of the focusing position. It follows that the change of focus caused by change of wavelength causes no trouble, provided that the position of focus remains the same for both specimen and reference material. The only exception to this is illustrated in Figure 11.2.

The determination of the correct focus is done by selecting by eye the position giving maximum image contrast at boundaries between specimens, or features, of different optical properties, and not that giving maximum measuring value from a homogeneous specimen. This is most conveniently done with white light or green light, for which the eye has its maximum sensitivity. At

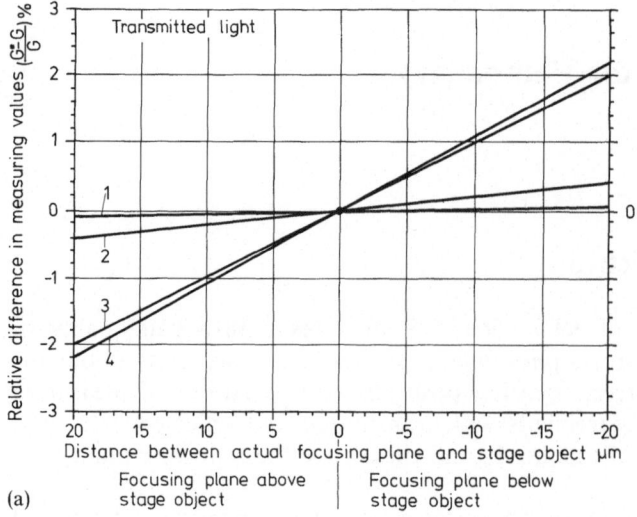

(a)

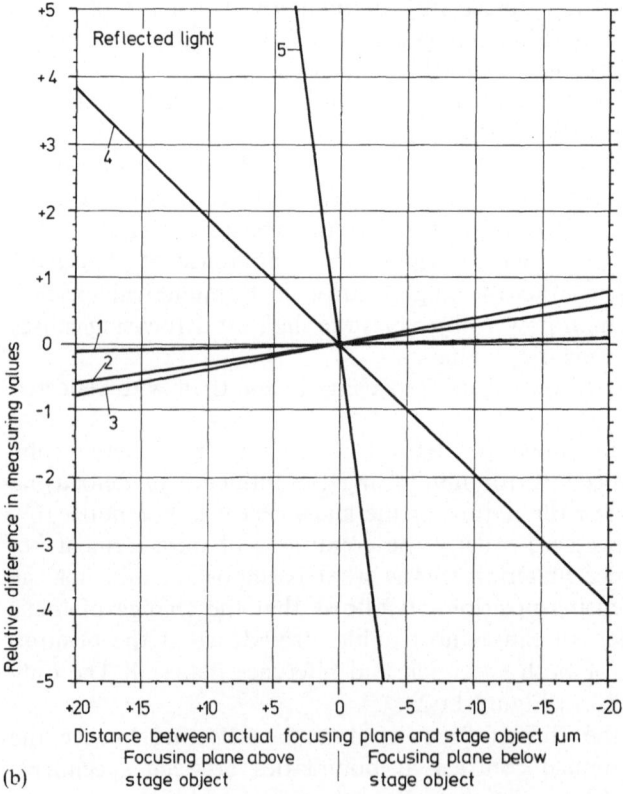

(b)

Fig. 11.1a and b. Legend see opposite page

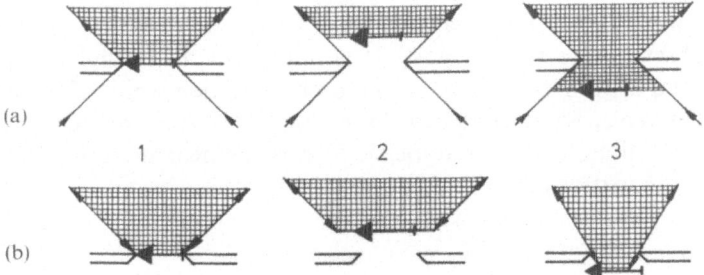

Fig. 11.2a and b. Encroachment of light from outside the area delimited by the photometer diaphragm when the stage object is out of focus (schematic).

(a) Transmitted light. (*a1*) Stage object in focus and filling the photometric field. (*a2*) Stage object situated above focus; light reaches the photocathode from surrounding areas. (*a3*) Stage object situated below focus; again light reaches the photocathode from surrounding areas.

(b) *Reflected light.* (*b1*) Stage object in focus and filling the photometric field. (*b2*) Stage object situated above focus, light reaches the photocathode from surrounding areas. (*b3*) Stage object situated below focus; unlike (*a3*) no light from surrounding areas reaches the photocathode.

Notes:

1. The level of correct focus is indicated by that of the back image of the photometer diaphragm.

2. All light within the shaded cone reaches the photocathode

Fig. 11.1a and b. Experimentally determined relative differences in measuring values for varying position of the focus of the objective.

Experimental conditions:

Graph no.		Objectives type planachromat		Effective numerical aperture of condenser	Diameter (in mm) of photometric field in the stage object
		Magnification number	Potential numerical aperture		
(a) Trans-	1	6.3	0.16	0.1	0.111
mitted	2	16	0.35	0.2	0.045
light	3	40	0.65	0.35	0.018
	4	100	1.25 (oil)	0.6	0.007
(b) Specularly	1	4	0.1	0.07	0.18
reflected	2	8	0.2	0.15	0.09
light	3	16	0.35	0.25	0.045
	4	40	0.85	0.55	0.018
	5	80	0.95	0.6	0.009

Note: The diameter of the luminous field was three times that of the photometric field. The relative difference (or relative error) is expressed by $[(G^* - G)/G] \cdot 100\%$.
G measuring value for perfect focusing; G^* measuring value for out-of-focus position of stage object.

Objects used for the tests: (a) Transmitted light; Homogeneous coloured gelatine foil (thickness 0.1 mm, refractive index 1.5) was placed between slide and cover glass and immersed in oil. (The transmittance of the foil was controlled by changing the wavelength and the tests were carried out for several transmittances). (b) Reflected light; Specularly reflecting surfaces of homogeneous objects such as mirrors or reflectance standards, with different reflectances.

Notes:

1. It was found that the absolute amount of transmittance or reflectance does not affect the result.

2. The measuring value at correct focusing position is not necessarily the maximum value

operation with UV or IR light, the image made visible with the help of an image converter is observed for this purpose. It is important to use light with the same spectral composition for focusing on both specimen and reference material. For measurements, we do not need to change the focus with change of wavelength, unless the limit of the photometric field is too near the boundary face of the structure to be studied.

11.2 Levelling of Stage Object in Specularly Reflected Light

A test for levelness is given in Chapter 8, Section 4 and here we invite attention to some practical points.

1. It is always necessary to level the area or the grain to be measured; it is not sufficient to level the general top surface of the preparation or its mount.

2. When the photometric field is near the edge between two objects of different hardness in a polished mount, the reflecting surface will be curved, even though the surface of the mount as a whole may be flat and levelled. It follows that such areas should be avoided as far as possible; if they have to be used, the actual area must itself be levelled and not just the object as a whole.

3. The specimen may have a surface ripple owing to uneven polishing of a structure within the object. The quality of the flatness of a surface is best observed by means of differential interference contrast (Chap. 6, Sect. 4.4, and Fig. 11.3c).

4. The levelling test must be carried out in such a way, that only light from the actual photometric field is allowed to contribute to the image of the condenser-aperture diaphragm. This can be achieved by reducing the diameter of the Bertrand diaphragm so that it selects just the required area in the stage object. Another way is to reduce the diameter of the photometer diaphragm to the required size and to view the image of the condenser-aperture diaphragm together with the image of the objective-aperture area that is formed at any level behind the photometer diaphragm. Such a level, for example, is that of the exit pupil of the photometer head (Fig. 7.1). A third way is to reduce the diameter of the luminous-field diaphragm to the exact size of the area to be measured. In such procedures it may well be necessary to reduce the opening of the diaphragms to pinhole size.

There are several auxiliary gadgets for levelling.

1. The polished specimen in its mount is placed on plastiline on a slide. The hand press is then used to squeeze the mount into the plastiline so that its upper face is made parallel with the lower face of the slide. Some elastic recovery of the plastiline may occur.

2. The auto-levelling stage (Fig. 9.5a). This is a metal plate with a hole inside through which the stage object is observed and measured. The lower face of the plate is pre-levelled so that after pressing the surface to be studied against this face, the surface is automatically levelled.

However, for microscope photometry it is necessary to ensure that the individual grain to be measured is itself levelled; this can be achieved by means of any of the following pieces of auxiliary apparatus.

3. The specimen mount rests either on elastic material or against a spring. Its upper face presses against the lower face of three disc-shaped screw heads, by means of which it can be levelled (Fig. 9.5b).

4. The specimen mount is fixed on a special slide having three screws. The surface of the specimen must be in a higher level than the screw heads so that the whole slide can be put in the hand press (see 1.) for preliminary levelling. The slide is put on the microscope stage; the screws then be used, if necessary, to correct the levelling.

5. The microscope stage is equipped with a mechanism that enables it to be tilted in all directions and the specimen with its mount is placed on such a stage.

11.3 Control of Photometric Field and Luminous Field

Size of Photometric Field. A photometric field of size sufficient to provide enough optical flux, thus radiant flux to give a reliable measuring value, is clearly requisite for obtaining a correct result. But particularly with biological material, many objects are homogeneous only over small areas. Hence, in order to keep the photometric field homogeneous, it must be adjusted to that area. The smallest diameter of this field in the stage object is 0.5 µm which is about twice the limit of resolution. This follows from the following argument:

With an immersion objective of $NA = 1.3$ and wavelength of 550 nm the limit of resolution is according to Eq. (2.1) 0.2 µm. If there are two separate points lying in this distance or whether the feature is a tiny dumbbell cannot be distinguished by the eye, because it is only the diffraction pattern that can be discerned and not the true geometrical image. For photometry the geometrical pattern is required, and this is obtainable only on a separation more than twice this distance; it follows that the lower limit for photometry is about 0.5 µm (see also Zimmer, 1973). This applies whether the photometer diaphragm is (theoretically) placed close to the stage object or, as it is in practice, in the plane of a real image. Consequently the diameter of the photometer diaphragm must at least be $(0.5 \, \mu m \cdot m)$, where m is the total magnification scale of the image at the diaphragm level.

Position of the Photometric Field. The photometric field must lie far enough away from interfaces of features (or structures) in order to avoid irregular propagation of light due to refraction, scatter, etc., at the interface. Of course, this demand cannot be fulfilled if the feature to be studied is too small, and in this case a certain measuring error must be accepted.

Diameter of Luminous Field. In order to suppress glare (Chap. 12, Sect. 3.1) the diameter of the luminous-field diaphragm has to be adjusted so that there

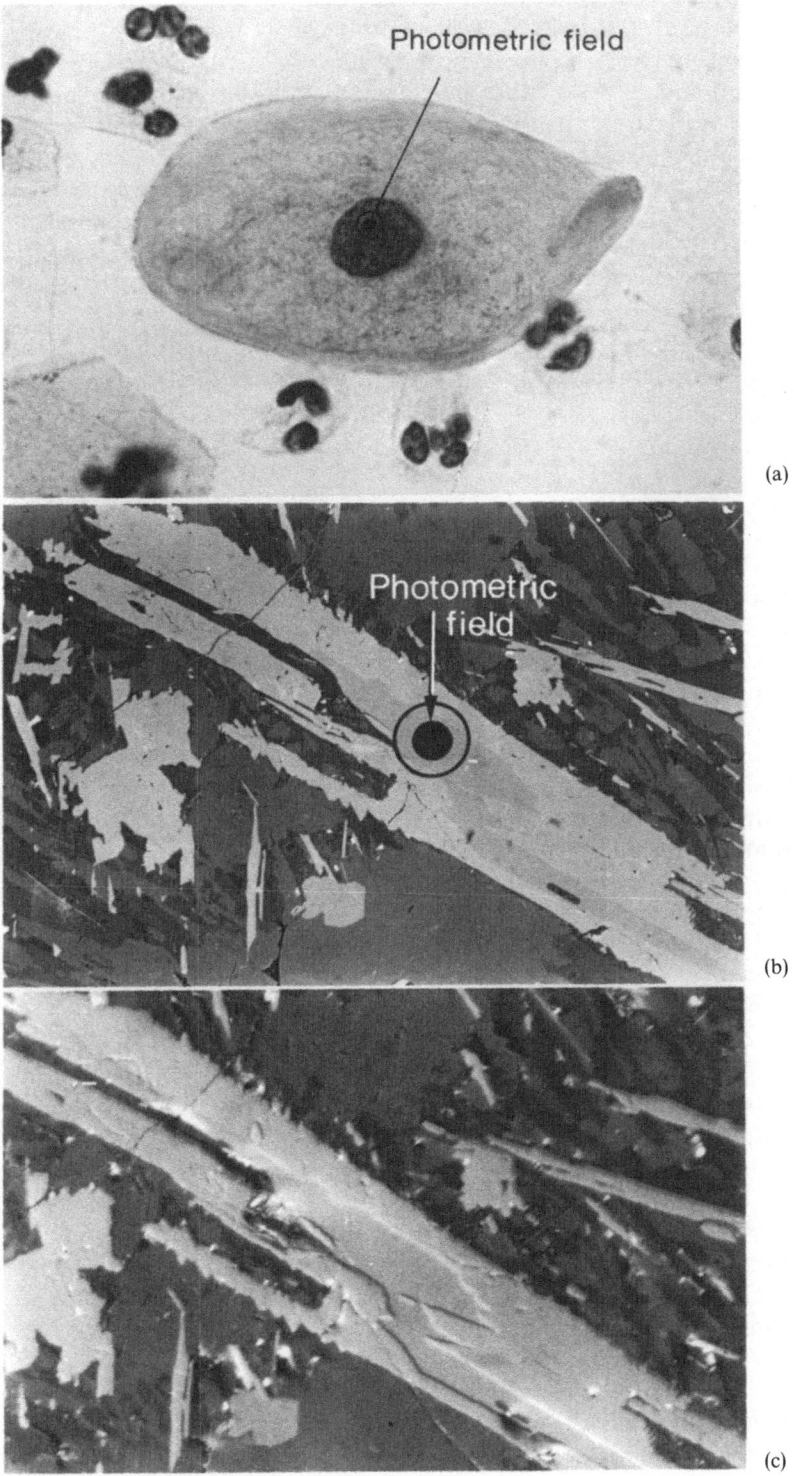

Fig. 11.3 a – c. Legend see opposite page

is no excess in illumination. But it should not be so small that the photometric field is shadowed or disturbed by diffraction fringes at the edge of the luminous-field diaphragm. In the rule, the diameter of the luminous field in the stage object should be twice to three times of the diameter of the photometric field as in Figure 11.3.

Centring of Photometric Field in Respect to the Rotation Axis of the Stage. For the comparison of measuring values for different orientations of the stage object it is important to maintain exactly the same area of the stage object within the photometric beam; this is required, for example in polarised light when the microscope stage is rotated. It follows that the photometer diaphragm has to have a circular opening and the rotation centre of the stage object has to lie exactly in the centre of the diaphragm. The centring procedure is described in Chapter 8, Section 1.

11.4 Control of Angle of Incidence

The measurements that we make, both of transmittance and of specular reflectance, are theoretically carried out at normal incidence. In Chapter 8, Section 4, we discussed the difficulties of approaching this conditions as regards the stage object with reflected light; in the present section we discuss the difficulty as regards the incident beam with transmitted and reflected light. It is possible only to approach the condition of perfect normal incidence and never possible to achieve it because at zero apperture angle no light would be seen at all. Thus the incident beam is always conical; if the cone angle is considered to increase, there must come a point beyond which the experimental conditions no longer correspond to the theory of optical properties of the stage object at normal incidence. Furthermore, in reflected light at operation with a prism reflector (Fig. 6.3) a certain obliquity of illumination must be accepted; this results in an increase of maximum angle of incidence for a given aperture angle. It follows that the maximum angle of incidence should always be measured at the start of any series of measurements of transmittance (absorbance) or reflectance (for the measuring procedure see App. 14); if necessary the angle should be systematically adjusted to the acceptable maximum size.

◀ Fig. 11.3a–c. Photomicrographs showing the adaption of the size of the photometer diaphragm and the luminous-field diaphragm to the size of the feature to be measured. The photometric field is indicated by a black circular disc; it is the photomicrograph of the actual opening of the photometer diaphragm. The luminous field is indicated by a circle; this is a drawing superposed on the photomicrograph. The luminous field was not photographed in order not to shadow the image.

(a) Photomicrograph of an epithel cell (contrast produced by Papanicolaou staining); transmitted light with planachromat 100/1,25 oil, magnification 1500:1. (b) Polished section of a coarse-grained lunar basalt; the dark area indicated is armalcolite. It is surrounded by ilmenite (brighter part of the grain). The reason for the restriction of the spot size is given in (c), where it is seen that armalcolite is harder than ilmenite. Specularly reflected light with planachromat 40/0.85, magnification 500:1. (c) The same section shown in differential-interference contrast which shows up the difference in level in the surface of the section due to difference in polishing hardness. The measuring spot must not contain differences in level

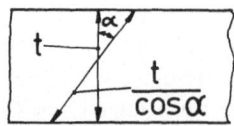

Absorbance (A₀) in the direction α=0, [A(α)] in the direction α=α

Fig. 11.4. Change in path in a parallel-sided plate for directions diverging from the normal.

The absorbance in the direction of the normal in the plate is expressed by $A_o = a \cdot t$; a linear absorption coefficient (for definition see App. 4); t thickness of the plate.

The absorbance in the direction α in the plate is expressed by

$$A(\alpha) = a \cdot t / \cos \alpha$$

so that

$$A(\alpha) = A_o / \cos \alpha \tag{11.1}$$

The effect of oblique incidence with transmitted light is described in the literature from the theoretical point of view (Blout et al., 1950). But the theory scarcely corresponds to the conditions in practice, so that we need not discuss the theory in this book. The only simple explanation that we can give in transmitted light is the following; in a parallel-sided plate the light path increases with increasing divergence (α) from the normal, because the light path is proportional to $1/\cos \alpha$; the same proportionality is true for the absorbance (A) of the plate provided that the light pencil is cylindrical and extremely thin (Fig. 11.4). Consequently the absorbance in the direction (α) is expressed by

$$A(\alpha) = A_0 / \cos \alpha. \tag{11.1}$$

A_0 absorbance in the direction ($\alpha = 0$). In practice, with normal bright-field illumination the light beam is not cylindrical but conical, the axis of the cone being the normal on the surfaces and the angle (α) being determined by the aperture angle (u) according to the law of refraction; $n = \sin u / \sin \alpha$ (n; refractive index of the plate), so that the relation between A and u is more complicate than expressed by Eq. (11.1).

The relation has been established experimentally by the author of this book using a clear coloured foil of given thickness for the test. The results of the tests are shown in Figure 11.5. From these we draw the following conclusions:

1. As predicted by theory, increase in numerical aperture produces an increase in absorption.

2. If we do not express absorption in terms of absorbance but in terms of transmittance the relation between absorption and divergence can, for the test material, be rather simply expressed approximately by

$$T(u) \approx T_0 \cdot b \cdot (1 - \sin u) \tag{11.2}$$

$T(u)$ internal transmittance of the plate for the aperture angle u (fraction)
T_0 internal transmittance for ($u = 0$)
b proportionality factor (fraction) [This factor depends on the refractive index of the plate (n), on the value of (T_0) and on the path in the direction ($u = 0$)]
u aperture angle

3. The less the transmittance (T_0) the greater the relative difference in transmittance for ($u = 0$) and ($u = u$) expressed by $[T(u)-T_0]/T_0$. If we take the difference

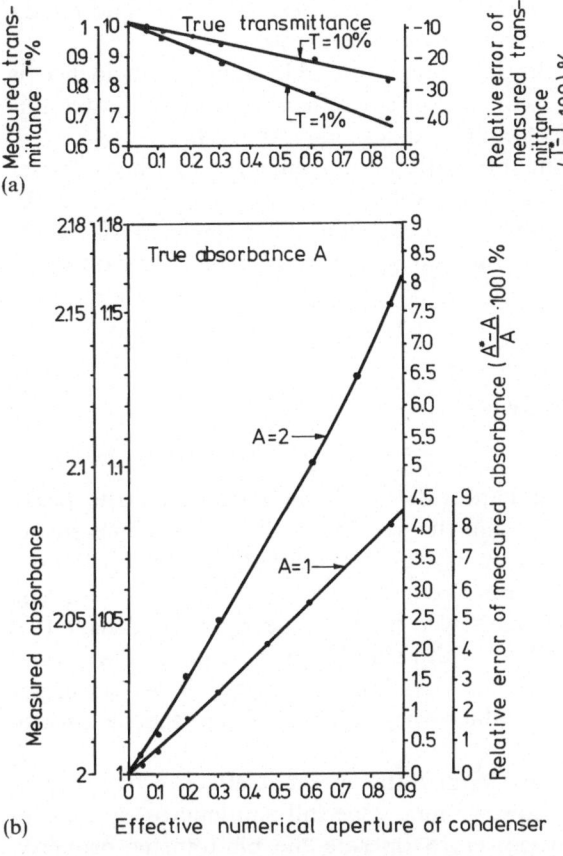

(a)

(b) Effective numerical aperture of condenser

Fig. 11.5. Differences in measured and true transmittance (or absorbance) due to excessive effective numerical aperture of the condenser.

(a) Transmittances; (b) absorbances corresponding to the transmittances in (a).

Operational conditions: Stage object: Homogeneous-coloured gelatine foil (thickness 0.1 mm, refractive index 1.5) placed between slide and cover glass and immersed in oil. Wavelength chosen so that the true transmittances of the foil were 1% and 10%. *Objective:* Achromat 40/0.85 oil. *Condenser:* Achromatic-aplanatic immersion condenser with potential numerical aperture of 1.4. The effective numerical aperture was varied between 0.01 and 0.85. *Reference material:* Spot on the slide outside the foil containing clear immersion oil. Diameter of photometric field in the stage object 20 μm; of luminous field 40 μm. *Measuring procedure:* For a given numerical aperture of the condenser the measuring value from the reference spot was set to 100% and the ratio of the measuring value from the foil and from the reference spot was taken as 'measured' transmittance. The ratio obtained with the smallest possible numerical aperture of condenser (for the present experiment it was $NA = 0.01$) was taken as true, hence as being (T_o). For each change of numerical aperture of condenser, the comparison of measuring values was repeated. The measured transmittances [$T(u)$] are plotted in Figure (a) and given on the left-hand ordinate. The right-hand ordinate gives the corresponding relative differences [$T(u) - T_o]/T_o$. Figure (b) shows the values of absorbance [$A(u)$] or relative difference of absorbance [$A(u) - A_o]/A_o$ derived from the measured transmittances

as being a systematic error and accept a maximum error of 5% the value of numerical aperture of illumination (sin u) must not exceed a certain maximum. This maximum is (sin $u = 0.3$), if ($T_0 = 0.1$), or it is (sin $u = 0.15$), if ($T_0 = 0.01$). With an oil-immersion objective and condenser the corresponding maximum values of numerical aperture are ($N \cdot \sin u = 0.45$) or ($N \cdot \sin u = 0.23$) respectively where $N = 1.5$; refractive index of immersion oil.

4. If we convert transmittance into absorbance and express the relative error in terms of absorbance [$A(u)-A_0]/A_0$ for a given numerical aperture of illumination the error in absorbance will be less than that in transmittance. If we accept a maximum error in absorbance of 5% then the maximum acceptable numerical apertures are (sin $u = 1$), if $A_0 = 1$ or (sin $u = 0.5$), if $A_0 = 2$.

Note: With globular stage objects there is no increase in path with increase in divergence and hence no error arises. With irregular stage objects the error cannot be estimated.

In reflected light tests were made using different test specimens and conditions, from which the following conclusions can be drawn:

1. With non-polarised light the maximum tolerable angle of incidence is 15° with the glass plate and 10° with the prism reflector; with the latter this comprises both the aperture and obliquity angles (Fig. App. 14b).

2. In linearly polarised light the international agreement is for the polariser vibration to be East-West; the maximum tolerable angle of incidence is 10° with the glass plate and 8° with the prism reflector. (The angle of 10° is the maximum angle of incidence when the full aperture area of a dry objective of numerical aperture of 0.17 is illuminated.)

11.5 Control of Spectral Bandwidth

The effect of too large a spectral bandwidth of light, is to lower the peaks and to fill the valleys in curves showing the spectral distribution of measuring values (Fig. 11.6) and hence of the property to be measured. Where it is desired to detect details in the shape of spectral curves (e.g. peaks, bands, steepness of flanks) the bandwidth must be narrower than the detail to be detected. It was found experimentally that the bandwidth should not be broader than about one tenth of the spectral width of the detail in the curve; if we want, for example, detect a spectral band of which the width is 50 nm, the bandwidth should not be larger than about 5 nm.

In Chapter 5, Section 4.3, we have studied the energetic effect of bandwidths and we learned that there is a lower limit, especially in microscope work, below which there is insufficient power to operate the photometer properly. If we use a prism- or a grating-monochromator, the radiant flux falls off with reduction of bandwidth as a second-order term, if the image of the exit is narrower than the aperture area or field; but it falls off as a third-order term if the width of the image is perfectly adjusted to that of the aperture area (or field), as shown in Figure 5.5 (aa) and (bd). It follows that in the latter case, it is the minimum geometrical width of the photometric beam at the monochromator exit that dictates the limit of radiant flux. It is found that this minimum is about 0.5 mm, so that with a strongly-dispersing monochromator a pass-band width as low as about 2.5 nm can be used; with the monochromator of more average design, the smallest bandwidth can be taken as about 5 nm.

With interference filters the bandwidth must be taken as about 15 nm for continuous filters. Homogeneous (plate) filters can, on the other hand be made to have a bandwidth below 5 nm (Table 5.1a); even with such a narrow bandwidth they allow a wide beam to pass.

Note: Attempts have been made recently to use tuned lasers to supply beams of sufficient radiant flux and highest spectral purity. But such highly-coherent beams suffer from interference on reflection from various glass surfaces in the optical train; this results in an unpredictable variation in the measuring values. If the coherence is artificially reduced the use of a laser may turn out to be beneficial. (See also note in Chapter 3, Section 4.2.).

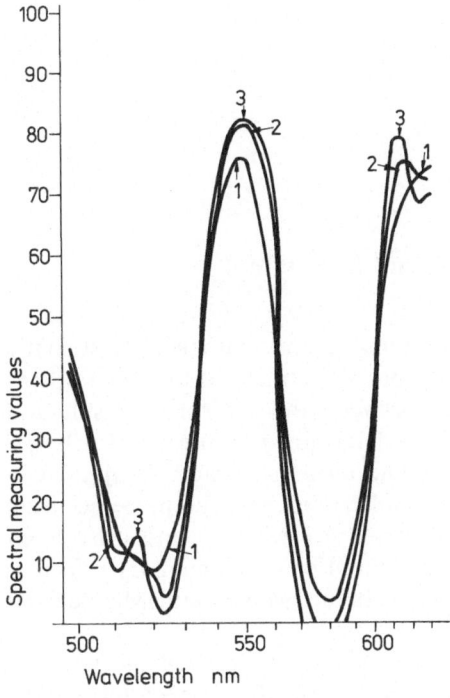

Fig. 11.6. Influence of the spectral bandwidth on the spectral resolution of measuring values. The curves were measured in transmitted light by the author under following experimental conditions:

Test specimen: Plate of didymium glass (Type FC 4, Schott, Mainz, FRG), thickness 2 mm. *Objective:* Planachromat 2.5/0.08. *Condenser:* Effective numerical aperture; 0.05. Diameter of the photometric field in the glass plate 8 mm, of luminous field 9 mm. *Monochromator:* Glass prism, type M 4 G II. *Recorder:* Type Servogor.

Curve no.	Bandwidths (in nm) for central wavelength (in nm)		
	500	550	630
1	12	17	25
2	6	8.5	12.5
3	1.2	1.7	2.6

It is clear that too large a bandwidth (a) lowers the peak value; (b) raises the minimum value; (c) obscures the detail of the curve.

Notes:
1. The curves are normalised to the value of 100% displayed with the empty field at the wavelength of 560 nm. For this kind of test the spectral behaviour of the measuring equipment need not be specified.

2. For this test material a reduction of the bandwidth below that of curve 3 produces no additional spectral detail

11.6 Setting of Central Wavelength

If the setting of the wavelength is not returned to exactly the same position in the method of wavelength change for test specimen and reference material (Chap. 9, Sect. 4.2) then an error will arise. The greater the spectral variation of measuring values the greater is this error. Steep ascents or slopes are extremely sensitive to this error source; they may be caused by pronounced spectral absorption or transmission bands in the stage object, emission peaks of the lamp or rapid change in spectral sensitivity of the photocathode near the limits of its spectral operating range.

The greatest stability in wavelength setting is provided by homogeneous interference filters. For hand-work a well-made click-stop is the best way to ensure a return to a calibrated position. Alternatively the scale to be set by eye must be sharp and easy to set accurately.

For the processing of spectral curves with the help of a computer, the strict repetition of wavelength setting and measuring-value input is required, while at operation with a recorder, a constant speed of movement and perfect synchronisation of movement of the recording paper with the movement of the wavelength mechanism is necessary.

CHAPTER 12

Systematic Measuring Errors

12.1 Distinction Between Systematic and Statistical Errors

There are two main categories of error: systematic and statistical. Systematic errors are due to defective apparatus or improper adjustment or incorrect operation; these faults give rise to a difference between the measuring result and the true value, as theoretically predicted or as determined by means of a better apparatus, properly adjusted and operated. The term 'accuracy' is applied in connection with systematic errors, and the smaller these are, the more accurate is the result. On the other hand statistical errors are random and arise from the scatter of measuring results about a mean value. For supplementary information on errors see Sandritter (1966); Goldstein (1970); Mayall and Mendelsohn (1970); David and Galbraith (1975).

In general each measurement is subject to errors of both categories, and so it is important to make the distinction and to use the correct terms. We call a measurement 'accurate' when it is free from systematic errors; we call a measurement 'precise' when the statistical errors are negligible. Precision can be increased by repetition of the measurements using the same apparatus, and the same procedure; precision is required for the detection of small changes in parameters. Accuracy cannot be increased in this way; accuracy is required in order to obtain verifiable data. A useful illustration of this difference is given in Figure 12.1. The present chapter is concerned only with systematic errors, and statistical errors are described in Chapter 13.

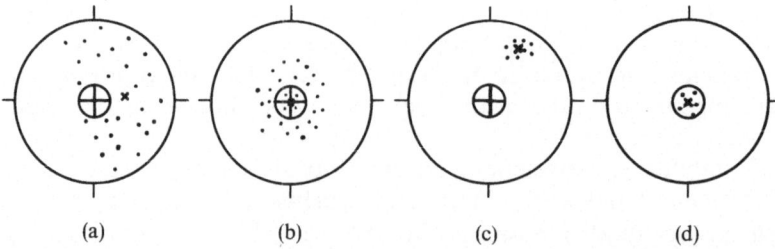

(a)   (b)   (c)   (d)

Fig. 12.1 a–d. Target in a shooting match illustrating the difference between systematic and statistical errors. x is mean value. The distance from the mean point to the centre is the systematic error. The scatter of the point is the statistical error; the term uncertainty is also used.
(a) Large statistical and large systematic error; thus low precision and low accuracy. (b) Large statistical and no systematic error; thus low precision and high accuracy. (c) Small statistical and large systematic error; thus high precision and low accuracy. (d) Small statistical and no systematic error; thus high precision and high accuracy

A distinction must also be made between errors in measuring (indicated) values and those in measuring results. Both statistical and systematic errors in measuring values have cumulative effects on the error in the measuring result. These effects can be described mathematically. An example is given in Appendix 3.

12.2 General Aspects of Systematic Errors

Systematic errors can be classified according to factors causing errors, and each type can be subdivided according to the particular fault; the contents of the present chapter covers important errors of this kind.

We have already stated that the magnitude of a systematic error is given by the difference between the measured (false) quantity (or the mean of a series of quantities) and the true (accurate) quantity; the false quantity may also be taken as 'error margin' or 'tolerance'. The true quantity is obtained by either adding or subtracting the difference from the measured quantity provided the difference can be calculated or else predicted. But, often, the amount of error is not known so that the true quantity cannot be determined at all. The error may be expressed absolutely, hence in the same unit as the quantity to which it is related, or as a fraction or multiple or percentage of the quantity. In the latter case it is expressed relatively.

As the various systematic errors combine in affecting the measurement, it is best to calibrate the apparatus by means of a series of stage objects of known properties covering its operating range (Exner and Schreiber, 1962). But it is important to determine the individual errors in order that steps may be taken to eliminate them, and we can list the various kinds of experimental procedure having this aim.

1. Use of calibrated reference material. A set of reference material of the same kind but containing various amounts of some measurable property (interaction factor) is tested. For example we can measure the reflectances of standards using first the lowest as the experimental 'standard,' and then repeating with the highest in this role. This is a very accurate procedure, and can be used to detect any error due to the secondary effect of glare in the microscope (App. 10); this accuracy applies only to the range of reflectance covered by the available reference materials.

2. Use of model specimens. Outside the range for which calibrated reference materials are available use can be made of model specimens; these are preparations, which can be of an ephemeral nature and of which the properties must be determinable by some method independent of the method in which they are being used as reference materials. For example a solution of known concentration of a dye of known absorptivity can be poured into a micro-cuvette of given thickness and the true absorbance calculated with the help of Eq. (9.7). Another type of model specimen consists of a plate of absorbing material; the true transmittance of this is measured with a macrophotometer.

Lastly model specimens can also be selected from test material having the same origin and nature as the specimen that is going to be measured. This can be done only when the behaviour of the test material can be predicted with accuracy. For example the use of erythrocytes and leucocytes is recommended as test material for measurements of DNA in biological specimens (Ruch, 1973).

3. Variation in the operating conditions of measurement. Any change in the measuring conditions has a complex effect on the measuring results. Here we give merely a general indication of the kind of effect produced by the main types of change, but reference can be made to the literature dealing with particular test material and to manufacturers' instructions. Later sections of the present chapter deal with the main types of errors and show what kind of operating conditions will minimise them. For example, in measurements of absorbance or in the determination of the amount of mass, it is necessary that the absorbance of the specimen should not be too great; this can be diminished either by reduction in thickness or by changing the wavelength. If the errors are connected with the spectral response of equipment, changing the photosensor or the lamp or the optical modules will affect the response of the whole system and thus change the amount of error.

4. Substitution by another type of apparatus. If the measuring results on a different apparatus are in close agreement with those obtained with the first apparatus, confidence in them is increased. If the same type of error occurs in both pieces of apparatus, this cannot be detected by this kind of test. Nevertheless an unknown apparatus should be calibrated against one which is known to behave well.

12.3 Instrumental Factors Affecting the Light Supply to the Photosensor

12.3.1 Glare

The term 'glare' is used for light that is unwanted in the image. In the literature it is sometimes referred to as 'Schwarzschild-Villiger effect' (Naora, 1952, Howling and Fitzgerald, 1959). It arises from reflection at optical interfaces or from scattering at lens mounts, and at impurities or dust in optical parts including the preparation. A significant amount of glare can arise from multiple reflection between the stage object and the front surface of the objective. In transmitted light an additional source of glare arises from multiple reflection between the upper surface of the condenser and the lower face of the specimen slide. These sources can be eliminated by operation with oil immersion objectives and by oiling the condenser to the specimen slide.

The amount of glare is constant for a given apparatus and for a particular adjustment of this; further, the darker the stage object, the more serious is the effect of glare; it causes diminution of image contrast to the viewer and in the photograph, while to the photometer it produces a measuring value additional to that of the stage object. The effect of glare is formally equivalent to the effect of other sources producing incremental measuring values, e.g. outside light, photosensor dark current, etc. For this reason, it is mathematically described in a general way in Appendix 5.

Glare may be observed in the centre of the image field manifesting a faint contour of a diaphragm or a lens mount, and this is sometimes observed with macrophotographs taken against a bright light source; it may also be observed in the microscope as a brightening more or less homogeneous, across the whole field. It may be pronounced in the transmitted-light image of a strongly absorbing stage object, in the reflected-light image of a weakly reflecting stage object, or in dark features of an interference image.

In measuring values the incremental value due to glare would be the same when the specimen is on the stage and when the reference material is on the stage. Except when the reference material and the specimen have exactly the same interaction factor, there will thus be a wrong measuring result, which is given by

$$\frac{IF_{sp}^*}{IF_{ref}^*} = \frac{G_{sp} + \Delta G_i}{G_{ref} + \Delta G_i} = \frac{G_{sp}^*}{G_{ref}^*} \tag{A.14}$$

IF interaction factor
G measuring value
sp   specimen
ref   reference material
ΔG_i incremental value

The asterisk indicates that the quantity is erroneous, because it is influenced by the glare.

Correction for Glare. The usual procedure is to correct for glare immediately in each measuring value. For this purpose, in transmitted light the whole apparatus is prepared for the measurement, care being taken, as far as possible, to exclude other sources which may produce incremental values. Then, the measuring value from a clear, perfectly transparent spot in the microscope slide is used as reference, and a perfectly absorbing feature is inserted in the photometric beam; the measuring value for this feature is that for glare (ΔG_i). (For the measurement of glare in images formed by interference see Chap. 10, Sect. 2.2.)

The measuring value for glare will be constant for a given set of operating conditions, and it must be subtracted from all subsequent measuring values for both specimen and reference material so that the corrected measuring values are the true ones. The glare correction is expressed by

$$G^* - \Delta G_i = G. \tag{12.1}$$

With bright-field reflected light glare is especially pronounced owing to reflection from the objective lenses and other glass surfaces that form part of the illuminating system between the reflector and the stage object (Piller, 1967). In this case, the incremental value for glare is that which is displayed when the stage object is completely removed from the photometric beam and a black box is placed in front of the objective as the stage object. In immersion, the front lens of the objective is dipped into a black box containing the oil (Fig. 9.3) or in a drop of oil placed on a plate of black glass that has the same refractive index as the oil. As an alternative to the immediate correction for glare as previously described, the error can be estimated by means of Eq. (A.16). But first of all every possible step should be taken to suppress glare, and the procedures for doing this are listed below.

1. Optics should be selected having the highest quality of anti-reflection coatings.

2. All glass surfaces in the optical train that are not necessary for the measurement should be removed.

3. The remaining glass surfaces should be perfectly clean.

4. All filter surfaces in the optical train should be slightly tilted so as to keep reflections out of the beam.

5. The illuminated area should be only slightly larger than the measured area (Fig. 11.3).

6. The aperture angle of the illuminating beam should not be the full aperture angle which the objective is able to accept, but should not be too small because this increases the relative effect of the glare.

Notes:

1. Use of a perfectly absorbing feature in transmitted light or the black box in reflected light gives the measuring result for glare normalised to the interaction factor of the reference material in the form

$$\frac{\Delta IF_i}{IF_{ref}} = \frac{\Delta G_i}{G_{ref}} = p \tag{A.15}$$

IF interaction factor
G measuring value
ΔG_i incremental value due to glare
ref   reference material
p short symbol for the ratio in eq A 15

The value of G_{ref} cannot be displayed in the test but only the sum of values $(G_{ref} + \Delta G_i = G^*_{ref})$. On the other hand, ΔG_i is displayed alone with the perfectly absorbing feature or the black box; thus G_{ref} is simply obtained by substraction of ΔG_i from G^*_{ref}.

2. In reflected light it may be beneficial to use a prism reflector rather than a glass-plate one simply because any reflection by a lens surface of the objective, so far the illuminating beam is involved, is directed back towards the lamp and not towards the photosensor; but this benefit will be obtained only if there is a properly adjusted condenser-aperture diaphragm (Fig. 8.2b).

Secondary Effect of Glare in Reflected Light. Previously we have stated that in bright-field reflected light the glare mainly arises by interreflections between lenses in the objective. But there are also inter-reflections between the lenses and the stage object surface. These cause an additional error; the measured reflectance of the specimen is too great or too small, even when a correction for glare, as previously described, is applied. The measured value is too great when the reflectance of the specimen is higher than that of reference material and too small for the inverse case. A short explanation is given in Appendix 10 from which we take the following mathematical correlation between the secondary-effect error and the factors involved in this error,

$$R_{sp}^{**} - R_{sp} = \Delta R_{sp} = R_{sp} \cdot \rho \, \frac{R_{sp} - R_{ref}}{1 - \rho R_{sp}}$$ (A.42)

R true reflectance (fraction)
sp   specimen
ref   reference material
ρ reflectance of the totality of glass faces lying between the reflector and the stage object (fraction)

The double asterisk indicates that the quantity is influenced by the secondary effect, although glare correction according to Eq. (12.1) has been applied.

Note: If it is necessary the absolute value of ρ is obtained as the measuring result of a test according to Eq. (A.15); it is expressed by

$$\rho = \frac{\Delta G_i}{G_{ref}} \cdot R_{ref} \, .$$ (12.2)

For a given set of objective and glass surfaces this value depends on the adjustment of the microscope.

Equation (A.42) makes it evident that the secondary-effect error is directly proportional to the reflectance of the surfaces lying between the reflector and the stage object (ρ) and to the difference in reflectances of specimen (R_{sp}) and reference material (R_{ref}); i.e. the smaller either of those factors, the smaller the error. It is obvious that no error occurs if the glass faces between the reflector and the stage object do not reflect or if the reference material has the same reflectance as the specimen.

On the other hand, experience has shown that the secondary-effect error may be distinct when ρ exceeds the amount of 0.05 (5%). It should be mentioned that with properly designed objectives and proper adjustment of diaphragms the value of (ρ) is usually much less than 0.05.

The error can be eliminated by one of the following procedures:

For a given objective with given adjustment of optics:

1. The diagram for graphical correction may be constructed by means of a set of three calibrated reference materials; black glass (NG 1), SiC and (W,Ti)C are very suitable. For the construction of such a diagram the reflectances of two of the reference materials, which, in this case, act as specimen, are measured with the third as the reference material. The measuring results are corrected for the primary effect of glare and are then plotted in a diagram against the two known reflectances. Then the measured points in the diagram are connected by a curve or a straight line as in Figure A.10 so that errors for measured reflectances of any amount can be interpolated or extrapolated. With the three materials three curves can be plotted, because each of the materials can be used as reference with each of the others as the specimen for the test.

2. The reflectance (ρ) may be measured according to Eq. (12.2) and the true reflectance of the specimen calculated after rearranging Eq. (A.38) to give

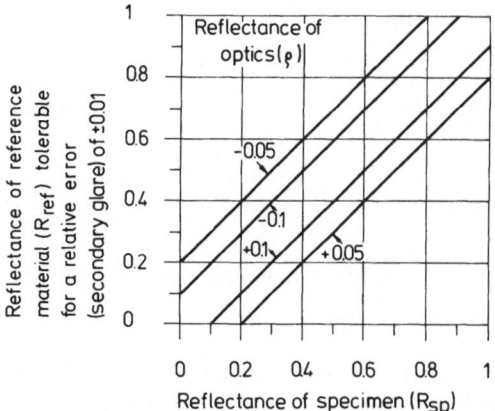

Fig. 12.2. Diagram showing the relation between the tolerable range of reflectance of the reference material that will keep the relative error in R_{sp} due to the secondary effect of glare to $\pm 1\%$. The graphs in the diagram are expressed by rearranging Eq. (A.42a) so that

$$R_{ref} = \frac{\dfrac{\Delta R_{sp}}{R_{sp}} (\rho \cdot R_{sp} - 1) + \rho \cdot R_{sp}}{\rho} \qquad (12.7)$$

R reflectance (fraction), sp specimen, ref reference material; $\Delta R_{sp}/R_{sp}$ relative error in the measured reflectance of the specimen due to the secondary effect of glare where $\Delta R_{sp} = R_{sp}^{**} - R_{sp}$; R_{sp}^{**} reflectance of the specimen containing the error due to the secondary effect of glare [The error is taken as being $\Delta R_{sp}/R_{sp} = \pm 0.01(1\%)$]; ρ reflectance of surfaces situated between the reflector and the stage object (The value of ρ is taken as being 0.05 or 0.1 respectively).

The graphs indicate the values of (R_{ref}) as ordinate values for varying values of (R_{sp}) as abscissa values; for a given value of (R_{sp}) the range of values of (R_{ref}) that will keep the relative error in (R_{sp}) in the range of $\pm 1\%$ is indicated by the interval of ordinate values on the normal to the abscissa value of (R_{sp}) between the points where the graph intersects the normal.

Examples for the use of the diagram: We take the true reflectance of the specimen as being $(R_{sp} = 0.3)$. If (ρ) is as large as 0.1 we must use a reference material the reflectance of which is within the range of $(R_{ref} = 0.2)$ and $(R_{ref} = 0.4)$ in order to keep the amount of $\Delta R_{sp}/R_{sp}$ equal to or less than 0.01. If (ρ) is as large as 0.05 (R_{ref}) can be down to 0.1 and up to 0.5

$$R_{sp} = \frac{R_{sp}^{**}}{1 + \rho (R_{sp}^{**} - R_{ref})} \qquad (12.3)$$

R reflectance (fraction)
sp   specimen
ref   reference material
ρ reflectance of glass faces lying between the reflector and the stage object (fraction)

The double asterisk indicates that the measuring result is influenced by the secondary-effect error although the correction according to (Eq. 12.1) has been applied.

3. The secondary effect of glare may be reduced to a negligible error by use of a reference material that does not differ in reflectance from the specimen by more than an accepted tolerance (Fig. 12.2). This is a simple way of avoiding this kind of error, but it makes great demands on the set of reference materials available.

12.3.2 Inter-Reflections in a Glass Plate Placed Obliquely in the Imaging Optical Train

Sometimes a glass plate with parallel faces is placed obliquely in the imaging optical train. This is done for two reasons:

1. The plate acts as a beam splitter and produces at the same time two images of the same stage object at different places. These images can be seen by two observers or else manipulated in different ways.

2. The plate acts as a reflector for reflected-light illumination (Fig. 6.1). In the latter case the splitting of the imaging beam is unwanted but cannot be avoided. In a normal reflected-light microscope only the imaging beam traversing the glass plate is of interest, while in an inverted reflected-light microscope either the transmitted or the reflected beam according to the arrangement of the optical system can be used.

This plate is sometimes situated in a sector that does not have parallel marginal rays, with the consequence that the imaging beam suffers multiple splitting and reflection in the glass plate; thus the regular image of any feature in the stage object is laterally displaced and has a superimposed parasitic image of an adjacent feature (Fig. 12.3b).

This parasitic image is formed almost at the same level at which the regular one is formed but with a certain lateral distance from the regular one. The distance depends on the thickness, refractive index and angle of inclination of the glass plate. For plates of standard design the distance is of the order of magnitude of one tenth to one millimetre in the plane of the primary image. The theoretical explanation of the formation of the parasitic images and the parameters involved in their geometry and intensity are given in Appendix 12.

Note: When the glass plate is situated in a sector of the imaging optical train having parallel marginal rays, and the image is formed with the help of a positive tube lens (Fig. 12.3a) the regular image is not displaced and no parasitic image is produced.

Of course, the parasitic image disturbes the clearness of the regular image both at observation and on a photomicrograph (Fig. 12.4). Further it causes an error in photometric measurements carried out just in the zone where the parasitic image is superposed on the regular image. In practice, the parasitic image can be seen only if it is formed on a dark background, i.e. if the regular image originates from a dark and the parasitic one from a bright area in the stage object. The visibility of the parasitic image can be reduced by properly coating the front and back face of the glass plate as explained below. But even with a coating the parasitic image may cause a serious measuring error. This error is the subject of the rest of this section. First we discuss its amount and then we describe the means for reducing it.

(a)   (b)

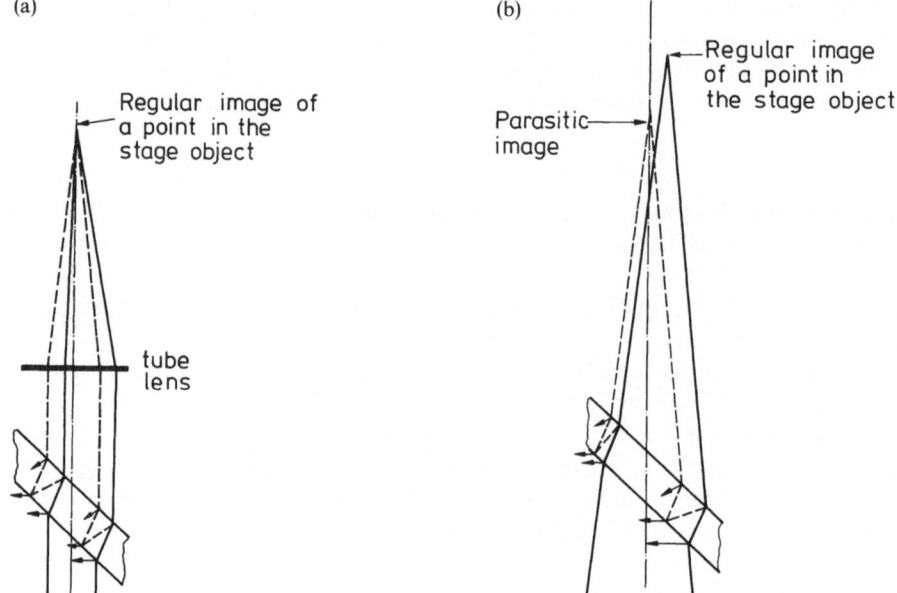

Fig. 12.3a and b. Effect of reflection and beam splitting in an inclined glass plate situated in the imaging optical train.

(a) Parallel marginal rays; (b) Convergent marginal rays. *Full straight lines:* regular rays (forming a regular image). *Dashed straight lines:* split-off rays (forming a parasitic image). *Arrows:* rays that need not be taken into account.

The figures show that there are reflections at the front and back face inside the glass plate and that these reflections result in the formation of a pencil of 'parasitic' marginal rays that is laterally displaced in respect to the pencil of the regular rays. In (a) the marginal rays incident on and traversing the glass plate are parallel [see also Fig. (2.1g and h)]. In order to form an image they must be made to intersect in the primary-image plane with the help of a tube lens. The tube lens causes both the regular and parasitic rays to intersect in the same point and hence there is no parasitic image. In (b) the marginal rays incident on and leaving the glass plate are converging towards the image plane. The parasitic rays intersect in a point which is to the side of and below the regular image point

Amount of Error Due to a Parasitic Image. We compare the measuring value for the area which contains both the regular image of the feature to be measured and the parasitic image of an adjacent feature with the measuring value for the reference material, selecting the measuring area in the reference material so that it is not superposed by the parasitic image of any adjacent feature in the reference material. However, the measuring area in the reference material is also superposed by a parasitic image but this image comes from the reference material itself. Under this condition the absolute error is expressed by

$$\frac{G_{sp}^*}{G_{ref}^*} - \frac{G_{sp}}{G_{ref}} = \Delta\left(\frac{G_{sp}}{G_{ref}}\right) = \frac{\rho_b \cdot \rho_f (G_2 - G_{sp})}{G_{ref}(1 + \rho_b \cdot \rho_f)}, \tag{A.58}$$

and the relative error by

$$
\frac{\varDelta\left(\dfrac{G_{sp}}{G_{ref}}\right)}{\dfrac{G_{sp}}{G_{ref}}} = \frac{\rho_b \cdot \rho_f (G_2 - G_{sp})}{G_{sp}(1 + \rho_b \cdot \rho_f)}
\tag{A.59}
$$

[For derivation of Eqs. (A.52) and (A.53) see App. 12]

G measuring value
sp   feature in the specimen to be measured
2   feature in the specimen adjacent to the feature to be measured
ρ reflectance of the glass plate (fraction)
b   back side (image side)
f   front side (stage-object side)
The asterisk indicates that the measuring value is influenced by light from a parasitic image.

From Eqs. (A.58) and (A.59) we draw the following conclusions:
1. With a glass plate of given reflectances on both sides and a given specimen, the error is directly proportional to the difference in brightness — expressed as difference in measuring value — between the feature in the specimen to be measured and the adjacent feature of which the parasitic image is superposed on the image of the first feature.
2. The error is positive, i.e., as previously stated, the interaction-factor measurement of the feature to be studied is too high when the adjacent feature is brighter than the first one; thus $(G_2 > G_{sp})$.
3. There is also an error when the adjacent feature is darker than the feature to be studied $(G_2 < G_{sp})$, although no parasitic image is seen in this case; this error is negative, i.e. the measured interaction factor of the feature is too low.
4. No error occurs when the feature to be studied and the adjacent feature have the same brightness; $(G_2 = G_{sp})$. This means that no error occurs with a homogeneous specimen.
5. For the beam traversing the glass plate the error is approximately proportional to the reflectance both of the front and of the back side of the glass plate. But since, in order to produce the image with the greatest possible brightness, one of the surfaces should have a reflectance of about 0.5 the opposite side should have a reflectance as low as possible in order to keep the error as low as possible. (In practice the surface on the stage-object side is that with higher reflectance.)

Suppression of the Effect of Parasitic Images. The effect of parasitic images is suppressed in several ways:
1. Use of a properly coated glass plate.
2. Select area in the specimen to be measured so that it is not too close to an edge but has a distance from the next edge of minimum amount d [according to Eq. (A.45)].

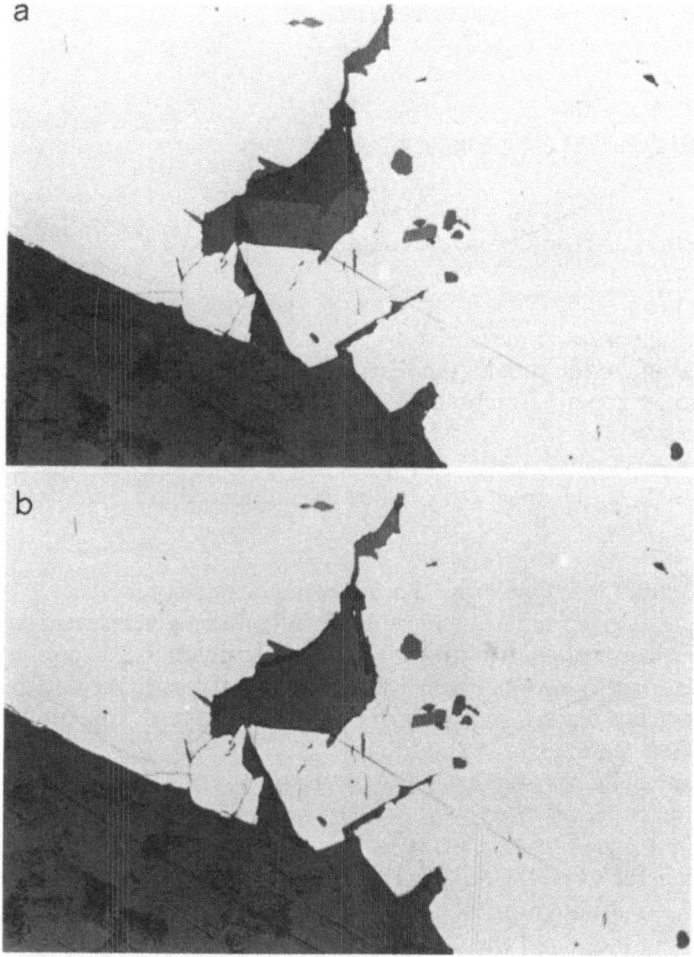

Fig. 12.4a and b. Formation of parasitic images by a glass-plate reflector in bright-field reflected light.

(a) Glass-plate reflector; (b) Prism reflector; (a) and (b) are photomicrographs of a polished section of arsenopyrite (*bright area*) and gangue mass (*dark areas*). Magnification scale 500:1.

The photomicrograph of (a) was taken with a glass-plate reflector. In the regular image of gangue mass (*centre*) the parasitic image of arsenopyrite is seen. *Small spots* with gangue mass at the *right* of this image are totally superposed by parasitic images so that the gangue mass appears brighter than it actually is. The photomicrograph of (b) was taken with a prism reflector. There are no parasitic images and the gangue mass is seen with correct brightness

3. Reduce the diameter of the luminous-field diaphragm so that the diameter of its image does not exceed the distance d in Eq. (A.45) and reduce the diameter of the photometer diaphragm correspondingly. (With this adjustment no parasitic image is formed because the feature adjacent to the area to be measured is not illuminated.)

4. Replace the glass plate, if possible, by another optical element that has the same function as the glass plate but does not produce parasitic images. Hence, in reflected light use a prism reflector instead of the glass-plate reflector (Fig. 12.4), or for the formation of two images of the same stage object at different places use a semi-reflecting layer cemented between the hypothenuse faces of two prisms so that the whole body is a cube. But with the latter system trouble can arise from glare due to reflections at the entrance and exit face of the glass body, unless these faces are properly coated.

12.3.3 The cos⁴-Law

The cos⁴-law concerns a pencil of rays having its axis oblique to the axis of the microscope. The volume occupied by this pencil is called light tube (Chap. 3, Sect. 4.1) and expressed in terms of optical flux. Hence this law expresses the modification of the optical flux in an oblique light tube in respect to that in a normal light tube; the optical flux in the oblique tube is diminished by the factor $(\cos^4\alpha)$ where α is the angle subtended between the axis of the microscope and the axis of the oblique tube. The derivation of this law is given in Appendix 13.

In practice the law plays a role for the optical flux of the sector in the optical train between the primary image and the exit pupil of the microscope, because in this sector α can have its greatest value. In this sector the effect of the cos⁴-law is that, for a light beam coming from the margin of the field of the primary image the optical flux is less than for a light beam coming from the centre of the field. The factor of diminution is $\cos^4(w/2)$, where $w/2 \equiv \alpha$ is half the field angle.

In an ocular of standard design $w/2$ can be up to $25°$ and so the factor of diminution can be as low as 0.66. In practice this factor can be even lower (down to about 0.5) because the law is not followed exactly owing to the complexities of the lens system in the microscope. Except in wide-angle oculars and objectives the eye does not detect this variation of intensity across the field, but it may be detected by means of the photocathode; this difference in detectability is due to the logarithmic response of the eye and of photographic emulsion, whereas the photocathode has a linear response.

Clearly this does not produce an error when the photometer diaphragm is fixed in the same way for both specimen and reference material; it is best to fix this at the centre of the field as this gives the greatest measuring value. But with beam switching in the image field (Chap. 9, Sect. 2) errors due to this cause can appear. For high performance neutral filters with gradual variation in transmittance in radial direction can be used to neutralize this effect, or it can be done electronically (balancing of background intensity from different points in the field), but such devices are costly and mainly used in television systems for electronic image analysis where they are indispensable.

Note: This source of error is the chief reason why a television system linked to the microscope and used for photometric work (image area scanning) cannot supply as high accuracy of the direct measuring result (Chap. 9, Sect. 5.2) as a photomultiplier that obtains the radiant flux through a fixed photometer diaphragm; in the television system measuring values from different places across the field are being compared and the natural differences in background brightness on these places cannot be perfectly compensated.

12.3.4 False Light from the Monochromatising Device

The origin and the effect of false light have been discussed in Chapter 5, Section 2. In the following we are dealing with the kind of photometric measurements for which the disturbing effect of false light is extremely pronounced. When the photometer is operated close to the limit of its spectral response, i.e. close to the limit of the spectral response of the photocathode or spectral transmittance of the optical system including the stage object, the measuring values, theoretically, approach zero. When the limit is exceeded, i.e. when the monochromatising device is set to supply a spectral band outside the limits, the measuring values, in practice, will not be zero but will have a certain magnitude due to the effect of false light.

For short, we call the range outside the limits the 'non-sensitive' range. For this range we obtain measuring results that are actually not existing but have the amount of the ratio of average interaction factors of the specimen and the reference material existing for the spectral range of the false light. (For theoretical explanation see App. 5.) The experience has shown that there may be false peaks or valleys in spectral curves constructed with measuring results from the non-sensitive range.

On the other hand, the display of measuring values in the non-sensitive range is evidence for the occurrence of false light, provided that there is no other source that may produce incremental measuring values, such as outside light or dark current.

False light is suppressed with the help of a pre-dispersing or blocking device as described in Chapter 5, Section 2; further, its effect is suppressed by chosing a photocathode which does not correspond to it, or we have to use a monochromatising device of better quality.

Note: Theoretically, the measuring results in the non-sensitive range due to false light would be distinguished from those due to other sources of incremental measuring values; the former results approach the mean value of the interaction factor of the specimen in the sensitive range or in the spectral range of false light and are influenced by the wavelength setting of the monochromatising device, the latter approach 1 and are not influenced by the wavelength setting. In practice, measuring results in the non-sensitive range arise from various sources so that no distinction is made.

12.3.5 Outside Light

Outside light is light that does not take the regular way through the microscope. Some light may enter through slits in tubes, housings or mounts or through any space, where the optical train is not closed, e.g. the space between the condenser and objective. Outside light may be natural room light, but even at work in a darkened room, it may come from the microscope lamp or from lamps attached to switches or scales (e.g. light-emitting diodes). Outside light causes incremental measuring values, of which the effect is described in Appendix 5.

In order to detect whether outside light disturbes the photometric measurement or not, the room must be darkened, the housing of the microscope lamp closed so that no light is emitted into the room, the microscope rendered light-tight and one has to observe any difference in the measuring value before and after the procedures. The amount of the difference gives us the incremental measuring value due to outside light.

Of course, the incremental value depends on the adjustment of optical and electric equipment including the setting of the monochromatising device (see also note on outside light in Sect. 3.4).

Outside-light effects can be suppressed by darkening the room and rendering the microscope light-tight. But if this, for any reason, is not possible, every measuring value must be corrected by subtracting the incremental measuring value due to outside light from every measuring value for the specimen and the reference material.

Note: Outside-light effects mainly occur with photometer systems that operate with direct light. In some makes of apparatus the light from the microscope lamp is modulated and the photoelectric assembly responds only to light of particular modulation, so that outside light not being modulated is disregarded (Chap. 7, Sect. 3).

12.4 Stage-Object Factors Affecting the Light Supply to the Photosensor

12.4.1 Features Disturbing the Light Beam

When the light beam hits a discontinuity in matter, its direction is changed in an unpredictable way; for transmitted light the effects are scattering, refraction, reflection, diffraction, and irregular absorption; for reflected light, the effects are scattering, irregular reflection, internal reflection, diffraction. The discontinuity may be natural or else caused during preparation of the specimen. It follows that light from outside the measured area on the stage object may hit the photocathode, while some part of the light coming from the measured area may be lost on the way. Such errors can never be eliminated completely, nor can they be described or determined quantitatively, and so they belong to the most important group of systematic errors. They can, however, be minimised by taking the following steps:

1. the photometric field is carefully selected for uniformity;

2. it is selected to be far from any disturbing feature, e.g. a boundary face;

3. the size of the luminous-field diaphragm is adjusted in relation to that of the photometric field (Fig. 11.3a, b).

But before the measurement, the specimen should be selected so that it clearly exhibits the property to be measured, and should be carefully prepared so that deformation and chemical reaction with agencies in its environment are kept to a minimum. The factors involved in the preparation of biological material are sampling, fixation, staining, cutting of thin sections, embedding and choice of embedding medium, which should have a refractive index as close as possible to that of the specimen and should be perfectly transparent. The use of the same preparation techniques is essential, if measuring results from specimens of the same kind must be compared. Hence the standardisation of techniques is beneficial.

Of course, careful preparation includes the use of a clean specimen slide, cover glass and embedding medium. It is also beneficial to filter the immersion oil before it is used in order to avoid deposition and migration of impurities in the photometric beam.

The preparation technique is also of a highest importance for the study of polished sections in reflected light. In these, in addition to the surface quality, the sub-surface condition may play a part. Cutting, grinding and polishing deforms the material, and metals have a very thin deformed layer. (Beilby, 1921; Rayleigh, 1937; Kranert and Raether, 1943). There is a rapid gradation over about 1 μm from the undamaged material to the finest crystalline, or amorphous, condition. The effect of this layer is complex; it may produce a measured reflectance that is too high, or too low; it is greatest with metals and alloys, less with ore minerals, and smallest of all with coals or transparent substances. But the possible effect due to the surface layer must always be kept in mind. This layer can be removed from metals or alloys by repeated alternative etching and polishing. Furthermore, there may be chemical reactions of the surface of the stage object with the preparation medium or the atmosphere.

It is at least necessary to keep the quality of the surface of specimens of the same kind equal. Hence the use of standard techniques in preparation of polished specimens is of extreme importance. Information on this may be obtained from the following literature: Schrader (1957); American Society for Metals (1958). (For the preparation of coal samples see ASTM D 2797.)

12.4.2 Irregular Distribution of Material

When the amount of material in a biological specimen is being measured by means of its absorption, it may not be possible to select the measured area to be smaller than the smallest homogeneous area in the specimen; in such case there will be a 'distributional' error (Ornstein, 1952). It is clear that this error will decrease with the decrease in size of the photometric field, but this relation becomes disturbed as the texture gets finer and as the diameter of the field in the stage object approaches the limit of resolution of the light microscope, so that the diffraction effects become important. However, in practice, the distributional error need be taken into account only if it is larger than the standard deviation of the measuring result due to overall statistical uncertainties.

The distributional error is due to the non-linear relationship between transmittance and absorbance. It is expressed in a relative term by

$$\frac{\Delta \bar{A}}{\bar{A}} = f(p, T_1) = \frac{p \cdot \lg T_1 - \lg[p(T_1 - 1) + 1]}{p \cdot \lg T_1} \tag{A.50}$$

$\Delta \bar{A} = \bar{A}^* - \bar{A}$ where

$\bar{A}^*$ false mean absorbance derived from the mean transmittance that is determined with a single measurement in the whole field of which a portion is occupied by material, the other portion is empty, hence perfectly transparent

$\bar{A}$ mean value of absorbance measured independently for the empty field and the material

T_1 transmittance of material (fraction)

p fraction of the field occupied by the material

[For derivation of Eq. (A50) see Appendix 11.]

The distributional error can be reduced in two ways:

1. By selecting the size of the photometric field as small as possible, e.g. to subdivide the area of the specimen into a mosaic of spots (areal scanning) and calculate the total mass of the chemical compound of the specimen in question from the sum of masses attributed to the single spots [Eq. (9.11)].

2. By applying the two-wavelengths procedure (Patau, 1952; Mendelsohn, 1966). This method can be used for material either stained or being self-absorbing. For this method, the photometric field is selected so that it is completely surrounded by a perfectly clear field. The measuring procedure involves the measurement of integral transmittance of material in the whole field at a wavelength λ_1. This must be a wavelength at which the Lambert-Beer law is rigorously applicable; then a second measurement at another wavelength λ_2, at which the absorbance of the same material is half of that at wavelength λ_1. For example, for Feulgen stain, λ_1 lies between about 565 nm and 580 nm, λ_2 between about 505 nm and 510 nm.

The measuring result is expressed by

$$m = \frac{F}{\varepsilon(\lambda_2) \cdot \ln 10} \cdot [1 - T(\lambda_2)] \cdot C \tag{12.4}$$

m mass of dissolved matter (in m mol)

F area of the photometric field in the specimen plane (in cm^2)

$\varepsilon(\lambda_2)$ molar absorptivity of material (in $cm^2 \cdot m\ mol^{-1}$) at wavelength λ_2, which is half of that at wavelength λ_1

$T(\lambda_2)$ transmittance at wavelength λ_2 (fraction) thus $[1 - T(\lambda_2)]$ absorptance

C a calculation constant defined by the following expression,

$$C = \frac{1}{2-Q} \cdot \ln \left(\frac{1}{Q-1} \right) \tag{12.5}$$

with the auxiliary quantity

$$Q = \frac{1 - T(\lambda_1)}{1 - T(\lambda_2)}. \tag{12.6}$$

Notes:

1. Because $T(\lambda_1) < T(\lambda_2)$, it follows that $Q > 1$.

2. The wavelengths (λ_1 and λ_2) are chosen so that the respective absorbances [$A(\lambda_2)$, $A(\lambda_1)$], or absorptivities, [$\varepsilon(\lambda_2)$, $\varepsilon(\lambda_1)$] have the ratio $1:2$, in the calculation, however, transmittances (T) or absorptances $(1 - T)$ are involved.

3. The wavelengths λ_1 and λ_2 are most accurately found by means of a test object or a series of test objects from the same slide that contains the specimen to be measured. The test objects are chosen so as to permit the most accurate determination of the absorbance at various wavelengths.

4. The two-wavelengths procedure is most conveniently be carried out with the help of an automated photometric equipment that controls the wavelength setting and processes the measuring values.

12.4.3 Failure of the Absorption Law

This topic is important for the study of biological specimens. A departure from the linear relation between concentration of material in the specimen and the measured absorbance can occur for reasons other than the distributional error. The Lambert-Beer law [Eq. (9.7)] can be applied for absorbances up to about 2 but, even for lower absorbances than this, the specimen has to be checked. It is necessary to ensure that there is no chemical reaction going on inside it or arising from interaction with light, or that the absorption band of a chemical compound (or dye-stuff) is not superposed by absorption bands of other compounds (or dye-stuffs). For the checking we need model or test materials of which the behaviour is predictable.

CHAPTER 13

Statistical Errors

13.1 Causes of Statistical Errors

Statistical errors, also called uncertainties, can be divided into four types.

1. Instrumental errors. These are caused by lamp instability, noise in the photosensor, and resistance fluctuation in electric circuits such as the power-supply units and photoelectric assembly.

2. Reading errors. These occur because the operator is not perfect in reading a fluctuating needle or digital display. In the case of on-line processing of measuring values with the help of a computer, or of print-out of measuring values, this type of error does not occur.

3. Stage-object errors. These arise because of natural variations in the stage object or else of artificial variations introduced by the preparation process. The first may in some cases be utilised in the characterisation of the stage object; the second must just be minimised by careful preparation.

4. Errors in repetition measurements. These arise whenever a module has to be re-set between measurements; they have to be minimised by reducing, as far as possible, the operations of re-setting and by using fiduciary marks or stops as aids in re-setting.

13.2 Expressions and Terms in Error Statistics

Statistical errors are ruled by the Gaussian law of statistical distribution, and the most important magnitudes involved in error statistics are described in the following. For more intensive information the reader is referred to the literature, e.g. Fisher, 1958; Ku, 1969; Hoel, 1971; British Standard, 2846; German Standard DIN, 1319.

The basic magnitude is the standard deviation expressed as

$$\sigma = \sqrt{\frac{\sum_{i=1}^{N}(x_i - \bar{x})^2}{N-1}}, \tag{13.1}$$

$$\bar{x} = \frac{\sum_{i=1}^{N} x_i}{N} \tag{13.2}$$

σ standard deviation in the unit attributed to the quantities x

x the single test value; it may be either a measuring value (G) or else a measuring result (M) (in the unit attributed to G or M)

$\bar{x}$ the arithmetic mean of all the test values

N the total number of test values (This number should be specified or indicated as subscript)

Note: The square of the standard deviation (σ^2) is called the variance, the relative standard deviation ($\sigma/\bar{x}$) is called the 'coefficient of variation'; it is usually expressed as fractional number or, more often, as percentage; thus ($\sigma/\bar{x}$) · 100%.

The standard deviation can be related to the probability (statistical certainty) with which we can be sure that the test values will lie between specified limits. Within the specified limits we can expect to find the following percentages of test values:

$(\bar{x} \pm \sigma)$, $(\bar{x} \pm 2\sigma)$, $(\bar{x} \pm 3\sigma)$

68.3%   95.4% (usually   99.7%
          rounded to 95%)

The range from $+3\sigma$ to -3σ is called the 'dispersion range'. In measuring techniques and in physical expressions, precision data are mostly referred to the standard deviation; in international constants and in biological expressions, it is the dispersion range that is mostly used.

Qualitatively we can say that the degree of precision varies inversely with the magnitude of the standard deviation, or with any magnitude related to this, e.g. the limit of confidence of the mean value (Sect. 3). It should be noted that the definitions given here apply for a sufficiently large number of test values and that, for such a number, the mean value is the same for all series of tests repeated under the same conditions.

13.3 Evaluation of Precision with a Limited Number of Tests

For a smaller number of tests of a series a spread of the mean value must be expected, and this can be calculated. The following definitions are applied to this spread:

Upper limit of confidence of the mean value: $\bar{x} + \sigma_N \cdot (f_N / \sqrt{N})$.

Lower limit of confidence: $\bar{x} - \sigma_N \cdot (f_N / \sqrt{N})$.

Range of confidence: $\pm \sigma_N \cdot (f_N / \sqrt{N})$.

Relative range of confidence: $\pm \sigma_N \cdot (f_N / \sqrt{N}) / \bar{x}_N$.

N number of tests
[For $\bar{x}_N$ see Eq. (13.2); for σ_N see Eq. (13.1).]

The factor ($f_N \geq 1$) is a calculation factor depending on N and on the certainty that is attributed to the confidence parameter; values of f_N are given in Table 13.1 along with values of $f_N / \sqrt{N}$.

Table 13.1. Values of the calculation factor (f_N) and of (f_N/N) under various conditions. (After German Standard DIN 1319)

Number of tests N	Statistical certainty					
	68.3%		95.0%		99.7%	
	f_N	$f_N/\sqrt{N}$	f_N	$f_N/\sqrt{N}$	f_N	$f_N/\sqrt{N}$
200	1.00	0.07	1.9_7	0.14	3.0_4	0.22
100	1.00	0.10	2.0	0.20	3.1	0.31
50	1.01	0.14	2.0	0.28	3.1_6	0.45
20	1.03	0.23	2.1	0.47	3.4	0.77
10	1.06	0.34	2.3	0.72	4.1	1.29
5	1.15	0.51	2.8	1.24	6.6	3.00

For most kinds of practical work N is selected between 10 and 100. When it is too small, the precision is low, and when it is too large the amount of time consumed in making the measurements is excessive as it achieves only a small gain in precision. For example, in order to restrict the range of confidence by a factor 3, N has to be multiplied by a factor 9 (see Table 13.1 and the quotients f/N).

For values of $N \geq 10$ the factor f_N is often neglected because it approaches 1, so that the confidence range of the mean value is simply expressed as

$$\frac{\sigma_N}{\sqrt{N}} = \sqrt{\frac{\sum_{i=1}^{N}(x_i - \bar{x})^2}{N(N-1)}}. \tag{13.3}$$

Of course, additional to the spread of the mean value, there is a spread of standard deviation depending on the value of N. But the spread of the standard deviation can be neglected for measurements with which this book is concerned. (If necessary for information on this kind of spread reference may be made to Fisher, 1958; Graf et al., 1966; British Standard, 2846.)

Note: The unqualified statement that the error of a series of measuring values or measuring results is ($\bar{x} \pm \Delta x$) is meaningless because the reader does not know if Δx is to be taken as σ, $\sigma/\sqrt{N}$ or some other precision quantity, or if it refers to accuracy and hence is to be taken as absolute or relative systematic error, error margin or tolerance. It is essential that what is stated should be clear.

13.4 Comparison of Precision in Different Tests

When two sets (1 and 2) of data obtained from the same number of measurements are available, we may wish to compare the respective precisions of these. For this we have a reference ratio which depends only on N and on the required statistical certainty (Table 13.2); this reference ratio is symbolised by Q_{ref}. We calculate the ratio of standard deviations of the two sets, using the larger

Table 13.2. Reference ratios (Q_{ref}) for comparison of the respective precisions of the two sets of measurements. (After Kaiser and Specker, 1956)

Number of tests N	Reference ratio (Q_{ref}) for statistical certainty of	
	95.00%	99.7%
∞	1.00	1.00
50	1.27	1.50
25	1.41	1.80
10	1.78	2.78
5	2.53	5.67

Note: These are the factors by which the standard deviation of a set of measuring values (1) must be larger than that of a set of values (2) before a difference in precision of the two sets can be asserted with the required confidence.

of both σ-values as numerator; thus $(\sigma_1/\sigma_2) \geq 1$, and compare this with the reference ratio. When the calculated ratio is less than the reference ratio a difference in the two precisions cannot be asserted, as they may, or may not, really differ. When the calculated ratio is greater than the reference ratio, the two precisions do differ, with the required statistical certainty.

It can be seen from Table 13.2 that, when N is small, the standard deviation of the set no. 1 must be considerably greater than that of set no. 2 before we can state that there is a difference between the precisions of sets no. 1 and no. 2; for example, if $N=5$ the standard deviation of set no. 1 must be 5.6 times that of set no. 2 in order to assert with a certainty of 99.7% a difference in the standard deviations.

13.5 Summation of Standard Deviations

In a statistical spread many kinds of errors are simultaneously involved, and so their effect is a summation of variances

$$\sigma^2 = \sigma_1^2 + \sigma_2^2 + \sigma_3^2 + \cdots + \sigma_n^2 \tag{13.4}$$

σ_1, σ_2 etc. standard deviations arising from source 1, 2, etc.

For example, the standard deviation due to instrumental noise can be obtained by processing the measuring values for noise with all the equipment left untouched. The standard deviation due to the stage object can be obtained by keeping instrumental noise to a minimum and then scanning the stage object; in such a test the scanning speed should be extremely low and the indicating module should be well damped, or else the mean of a sufficient large series of values of the same spot (Chap. 7, Sect. 3.1) should be taken as the actual measuring value.

Since the integral standard deviation is determined by the sum of squared terms, only terms of greatest magnitude need be taken into consideration, the contribution of smaller terms being negligible. If, for example, $\sigma^2 = \sigma_1^2 + \sigma_2^2$ and σ_2 is one half of σ_1; the value of σ is only $1.12 \cdot \sigma_1$; hence the source (2) contributes only 12% to the total standard deviation.

The summation of variances should be distinguished from their accumulation, which is due to errors of individual terms in an equation, being used for the calculation of the measuring result. Accumulation is explained by means of the differential calculus. An example is shown in Appendix 3.

CHAPTER 14

Application of Microscope Photometry

14.1 Absorption Parameters Measured in Transmitted Light

14.1.1 Study of Biological Material

In the photometric study of biological material it is mainly the measurement of absorbance that is involved, because this quantity is a linear factor in Lambert-Beer's law [Eq. (9.7)]. Other absorption parameters are applied in rarer cases, but unfortunately, there is a confusion about definitions and names for these (see App. 2.4), so that care must be exercised in making a comparison of data published by different authors.

There are three topics:
1. The determination of the amount of mass of certain constituents, either absolutely or relatively, and of local variation in this; alternatively the concentration can be measured.
2. The identification of constituents in specimens, e.g. in forensic work.
3. The recognition of unusual optical or geometric properties distinguishing some specimens from others that seem at first to be of the same kind.

Topic (1) involves finding the wavelength for which the substance is most absorbing, either stained or unstained. The mass can then be calculated by means of Eq. (9.10). The area occupied by the absorbing constituent and its molar absorptivity are used as constants in the calculation. When it is required to find the amount of mass in all the measured spots, the products of absorbance and area are summed by means of Eqs. (9.11) and (9.12); areal-scanning procedures are very useful in this.

In calculating the concentration we use Eq. (9.8); the molar absorptivity of the constituent and the thickness in the direction of the microscope axis are used as constants. The constituents measured in this way include: nucleic acids (DNA and RNA), proteins, aromatic amino acids, hemic compounds, nucleotides, enzymes and many other chemical components of biological bodies. (Naora, 1958; Walker, 1958; Pollister and Ornstein, 1959; Grundmann et al., 1963; Deitsch, 1966; Rudkin, 1966; Sandritter, 1966; Freed, 1969; Caspersson and Lomakka, 1970; Green, 1970).

Topic (2) is a generalisation of topic (1) in which the spectral absorption as a whole is measured for comparison with that of other substances. When only the relative spectral distribution of absorption, or the spectral position of absorption peaks or valleys, is required, the thickness need not be taken into account, but when absolute values are required for absorption parameters we have to know or else to determine the molar absorptivity as we measure the absorbance, we need to know the thickness and concentration of absorbing material (Burnett, 1972).

Topic (3) involves the comparison of local variations of absorbances at particular wavelengths, as for example in the discrimination between normal and abnormal specimens or populations of specimens. In such studies a line- or areal-scanning procedure is always used. This kind of study is very important in diagnosis of malignant tumours or unusual chromosomes.

Note: The substances that are formed in biological processes may be either self-absorbing or rendered absorbing by means of dyes. Self-absorption mainly occurs in the UV region as in purines and puridines (260 nm), the second peak at 280 nm in aromatic amino acids, and the 340 nm absorption band in nucleotides. Hemic compounds have absorption bands between 400 and 450 nm, while most dyes absorb in the visible range.

14.1.2 Study of Non-Biological Material

Some of what has been said in the previous section applies well here, but reflected-light measurements play a larger part in non-biological materials than do transmitted light measurements. Reflected-light measurements are used where the absorption is too strong for the use of transmitted light. Again there are three topics:

1. Simple identification.
2. Partial information about chemical composition in a known series.
3. Information about chemical bonding in the crystal structure.

Topic (1) concerns materials that are generally more homogeneous than biological specimens: they are mostly crystalline, although glasses and fibres (plastics) are included. Minerals and rocks have been prepared in thin sections for more than a century, these being embedded and mounted on glass slides. Polarised light is always used, and the optical properties of minerals that are effectively transparent in a thickness of 0.03 mm have long been used by petrographers. Single transparent bodies, e.g. crystals can also be mounted for optical study in transmitted light, and this is done for artificial inorganic and organic substances as well as for minerals.

Photometry enables us to establish the characteristic absorption spectra of coloured substances, and even if no coloration effect is visually observed, the substance may have specific spectral distribution of absorption in the UV or near IR range that can be measured. In optically anisotropic material absorption varies with vibration direction of linearly polarised light in the crystal (pleochroism). Measurement of absorption parameters is an excellent supplement to the determination of refractive index, which is the basic optical constant for identification (Weigel and Ufer, 1927; Mandarino, 1959; Cheznokov, 1960; Toubeau, 1961; Adams, 1965; Faye and Nickel, 1969).

The application of this procedure for detecting evidence in forensic criminal techniques will be specially mentioned here (Frei-Sulzer, 1965).

Topic (2) concerns materials in which one main constituent chemical element can be replaced progressively by another as, for example, an isomorphic series in minerals. The change of chemical elements may result in a change of the

absorption spectrum that permits determination of the composition by photometric measurements. This applies in practice only to simple replacements,
since interference by additional replacements usually makes the change too
complex to interpret (Lehmann and Harder, 1970).

Topic (3). A few studies have made use of features in the spectral curves
of optic properties (refraction, absorption) to derive information about the
inter-atomic bonding in the crystal structure. The recent improvements in apparatus and methods make it likely that more use will be made of this source
of information, especially as the method lends itself to extension into the UV
and the IR regions of the spectrum (Yamada and Tsushida, 1956; Anex, 1966;
Burns and Vaughan, 1970; Preuss, 1973; Langer and Abu-Eid, 1977).

14.1.3 The Utilization of Photographs

Since in Chapter 10, Section 5 the photometric measurements on photographs
(micro-densitometry) has been described, here we deal only with the specimens
for which this kind of measurement is used. The utilisation of photographs
implies, above all, a regular relationship between the blackening in the photographic negative (in some cases it can be the positive that is used) and the
quantity of light that has been emitted (transmitted or reflected) by the object
on the photograph.

Photographs can be divided into three categories:
1. Photomicrographs,
2. Macrophotographs,
3. Photographic emulsions having a certain blackening pattern.

Category (1) contains photographs of specimens that cannot, for some reason, be measured directly under the microscope. This occurs for example when
the specimen is so small that the required feature cannot be isolated by the
photometer diaphragm. It occurs, too, whith genes in chromosomes, of which
the absolute absorbance is to be measured, or of which the arrangement in
the chromosome is indicated by the distribution of the absorption. It occurs
also with features, that cannot be exposed to light or don't absorb light, the
photograph is used instead to show the effect of X or electron radiation on
the specimens such as objects photographed with the electron microscope (Bahr,
1973). Finally micro-autoradiographs in films stripped off from the radioactive
material belong to this category.

Category (2) contains photographs of macroscopic objects, which are to
be classified according to 'brightness' or 'darkness' or by patterns of bright
or dark spots. Aerial photographs taken from the vegetation on the earth's
surface provide an interesting example; diseased vegetation can be noted through
changes in the blackening on the photographs as compared with those of healthy
vegetation. In such investigations colour photographs are also used since changes
in colour can show up the same effects (Jackson et al., 1971).

In category (3) are found photographic materials showing spectral lines
photographed through a spectrograph, or diffraction patterns obtained by means
of any kind of diffraction apparatus.

Before the advent of modern X-ray spectrographs, micro-densitometry on X-ray photographs was very much used, especially for powder photographs (Herz, 1955).

Note: By moving the photograph through the photometric beam (areal scanning, Chap. 9, Sect. 6.2 the distribution of measured spots across a certain area, and hence area coordinates for the measured features, can be determined.

14.2 Parameters Measured in Reflected Light

14.2.1 Specular Reflectance at Normal Incidence

Specular reflectance serves to determine the basic optical constant (refractive index) of transparent material, that cannot be prepared in the form of a thin section on a microscope slide for normal visual inspection in transmitted light (see Chap. 9, Sect. 5.4). In such cases Eq. (9.17) is used for the derivation of the refractive index from the measured reflectance.

Note: The measurement is disturbed by internal reflections and reflections at the lower face and side faces. The effect of these reflections can be reduced by embedding the specimen in a liquid with similar refractive index or using an objective with short focal depth.

Also substances that are too strongly absorbing to be studied by transmitted light, have to be studied in reflected light. Since two basic optical constants (refractive index and absorption coefficient) are involved in the reflectance, the latter has to be measured in air and in another medium to give the two equations required for the separate derivation of the constants according to Eqs. (9.14) and (9.15). While oil is used as the second medium mostly for the study of ore minerals and coals, an interference layer (Chap. 10, Sect. 6) has proved to be advantageous for the study of metals and alloys. For mere determinative work it is mostly sufficient to know only the reflectances in air and in oil. Hence we measure the reflectance for a given wavelength and/or its spectral distribution; in an anisotropic section we measure these quantities for both vibration directions and hence derive the bireflectance. With 'white' light the spectral distribution of reflectances produces colour (and pleochroism) and this is a useful property in the identification of certain minerals, especially now that the colour can be defined quantitatively (Piller, 1966; Henry and Phillips, 1977).

In Ore Mineralogy reflectance values were introduced into diagnostic tables some decades ago (see literature references by Ramdohr, 1975) but the generalisation of this has had to wait for the development of modern apparatus and techniques. It is only now that this can be considered to have been achieved with the publication of the IMA/COM Determinative Tables. Modern Polishing techniques and the prevision of good international standards have played an essential part in this development. (See also Koritnig, 1964; Cervelle and Lévy, 1971; Uytenbogaardt and Burke, 1971; Galopin and Henry, 1972; Bessmertnaja et al., 1973; Vyalsov, 1973; Bowie et al., 1975.)

In Coal Petrology, reflectance is used to distinguish between the constituents of coals called macerals and to determine their 'rank' (degree of coalification) which is the alteration of the original peaty constituents into coal material through the stages of bitumineous coal. The process involves loss of volatiles and the increase of carbon content; anthracite is the highest stage of economically valuable coal and the end product is graphite; reflectance increases as the carbon content increases and the volatile content decreases (Mackowsky, 1960; Gilbert, 1961; Murchison, 1964; American Standard ASTM D 279). The reflectances met with in these studies are all rather low and they lie within a small range.

In practice, it is mainly the reflectance of the maceral group of vitrinite that is measured, and this is done in oil; the range of values being up to 4%.

Much work has been done in the reflection behaviour of coal in recent years, such as the determination of rank and composition since this information helps in the study of the structure and genesis of coals. But little use has been made so far of the correlation between spectral distribution of reflectance and of coal chemical composition, although this offers a possible means of obtaining information on the formation of aromatic-type molecules in coals.

Note: Some people believe that the low reflectance of the constituents of coal requires a higher accuracy in the measuring result than does the higher reflectance of ore minerals or metals, but this is a mistake. Of course, as high accuracy and precision (Chapter 12, Section 1) are desirable for measurements on all kinds of material. The highest degree of accuracy and precision that can be achieved depends of the quality of measuring equipment and on the skill of the operator in suppressing sources for errors, and this is the same for all kinds of materials.

What, however, is actually required for measurements on coal is the adequate sensitivity of the photoelectric assembly (see definition of sensitivity in Chap. 3, Sect. 7). This demand is explained as follows: We take the scale of the indicating device to contain 1000 intervals (or measuring-value units), so that the maximum measuring value is 1000. The decimal points are not important so that the maximum value can be as well 100.0, 10.00, etc. At first, we take the photoelectric assembly to be adjusted so that reflectances of 0 to 100% are spread across the whole scale; the sensitivity being one measuring-value unit per 0.1% reflectance.

For the measurements on coal only reflectances of, say, 0 to 10% or 0 to 1% need be taken in consideration. We adjust the photoelectric assembly so that maximum measuring value corresponds to 10% or 1% reflectance, hence the sensitivity is increased up to one measuring-value unit per 0.01% or 0.001% reflectance respectively.

In artificial minerals, such as ceramics and cements, and in metals the conditions of measurement are similar to those in ore mineralogy, but such measurements are still far too little used (Trojer, 1962; Knosp, 1970; Hummel, 1971). It is again a matter of observers not using the method because of lack of tables, and because they do not make measurements, tables cannot be compiled.

14.2.2 Diffuse Reflectance from Surfaces

Diffuse reflectance is measured from unpolished flat surfaces. The measuring result serves for specifying the following properties: quality (roughness, cleanness, homogeneity), colour, and local or temporal variations of these (Peltz et al., 1973). The measurements are also used for the microscale of accumulations of organic material due to an electrophoretic process across a given substrate (Edström, 1973; Neuhoff, 1973). The material to be investigated may be paper, wood, metal, or plastics.

Areal scanning is often used, and so many spots can be measured on the surface. For example in the quality control of paper on which printed figures are to be processed electronically, the mean diffuse reflectance and the standard deviation are required so that impurities may be detected.

14.2.3 Reflectance from Micro-Autoradiographs

The measurement of the reflectance of silver grains in micro-autoradiographs is of great importance in biology. For irradiation by beta rays emulsions are used that have small, uniform and very regular grains. When the specimens are dispersed elements such as cells and tissues etc., the emulsion is smeared across the preparation and, after exposure, developed without being stripped off. Certain components of the specimen have previously been labelled with radio-isotopes, for example blood cells with ^{14}O and DNS with ^{32}P (Pelc, 1958; Dörmer, 1973).

After development the blackened silver grains permit localisation of the emitting particles, while photometry yields a measure of the concentration (relative or absolute) of the labelled components; areal distribution is also obtained. Of course all such measurements are valuable only when there exists a regular relationship between the measuring value and the number of silver grains on the one hand and on the other, between the number of silver grains and the quantity of the labelled component. (For the procedure of measurement see Chap. 10, Sect. 4.)

14.3 Microscope Fluorometry

Microscope fluorometry involves the measurement of fluorescent light. The procedure is described in Chapter 10, Section 1. It is mainly used for the quantitative determination of certain constituents in biological materials, such as nucleic acids, amino acids, and proteins, which are either self-fluorescent or become so when stained with fluorochromes which react stoichiometrically with the constituent in question (Mansberg and Kusnetz, 1966; Konev, 1967; Witte, 1975; Ruch and Leemann, 1977). The relation between the fluorescent intensity and the amount of the constituent is taken to be linear; otherwise a calibration curve is first established by means of standards (Chap. 9, Sect. 3.4). The actual measurement is quite easy because a single result from the entire specimen gives the total amount, whether the constituent is homogeneously distributed, or not. In this, it is unlike the determination of mass by means of a series of absorbances.

In recent years this type of measurement has been applied to fossil plant material and to organic matter in rocks in order to determine the degree of coalification, etc. Fluorescence in coal-like materials changes inversely with the degree of coalification while specular reflectance changes directly with the degree of coalification; thus the measurement is made chiefly on brown coals and peats. The constituents are self-fluorescent, and both the intensity and the spectral distribution of the fluorescence are of interest. Information on the diagenetic processes over geological periods can be obtained from such data (van Gijzel, 1973; Ottenjann et al., 1975).

Similar studies have proved useful also in the identification of certain mineral substances present in small amount in rocks; gemstones too can be usefully studied in this way (Monés Roberdeau et al., 1975). In forensic studies also the method is increasingly used, but so far only qualitatively; refinement by use of quantitative measurements is a step likely to be taken in the near future.

14.4 Refraction Parameters Derived from Optical-Path Difference

In Chapter 10, Section 2 we learned the use of photometry for determination of optical-path difference and the formula for calculation of refraction parameters correlated to path differences. The present section briefly describes the kind of materials for which refraction parameters are important.

The refractive index is used for the identification of solid transparent materials on the microscale, e.g. crystals, glass bodies, fibres, plastics, etc. For tiny grains this is sometimes the only parameter that can be used for identification, unless X-ray analysis is applied. The corresponding determinative tables are found in the relevant literature, specially in that dealing with minerals. It should be mentioned that, as opposed to immersion methods which are based on the equating of the refractive index of an embedding liquid with that of the unknown specimen, the derivation of refractive index from optical-path difference does not require the procedure of continuous or stepwise change of the refractive index of the embedding medium; further, it allows the determination of refractive indexes that are greater than the maximum of the available embedding media.

Microscale thicknesses can be measured in this way, such as epitaxic layers on a substrate, crystal growth in steps, etc. In biological materials, the method is used for determination of the thickness of thin layers of transparent substances. Concentration, or mass, of components in cells, tissues, chromosomes, etc. are measured in this way on unstained material. The technique requires effectively transparent substances, and so is complementary to the measurement of absorption in more or less absorbing substances (Davies, 1958; Beneke, 1966).

Appendix

1. Specification of a Light Wave

Briefly stated the concept of light in this book is applied to electromagnetic waves in the spectral range 250 nm to 1100 nm, and this range is divided into the UV range; 250 to 400, visible range; 400 to 700, and near IR range; 700 to 1100. This classification is chosen for practical reasons. There are several ways in which a light wave may be defined, and these are all inter-related, as shown in the following compilation:

a) λ wavelength (in nm $= 10^{-7}$ cm); the distance between successive points having the same phase along the wave normal

b) ω angular frequency (in rad$\cdot$s^{-1}); the angle turned through by the wave vector in 1 second

c) ν frequency (in s^{-1}); the number of waves passing a point in 1 second

d) E_{ph} photon (in eV), the energy of light waves expressed in electronvolts

e) $\bar{\nu}$ wave number (in cm^{-1}); the number of waves in a path of 1 cm in vacuo

The formulas of inter-relation are:

$$\omega = \nu \cdot 2\pi$$

$$\nu = c/\lambda$$

$$E_{ph} = h \cdot \nu = \frac{h \cdot c}{\lambda} = \frac{1239.7 \text{ nm}}{\lambda}$$

$c = 2.998 \cdot 10^8$ m$\cdot$s^{-1} the speed of light in vacuo

$h = 6.62_5 \cdot 10^{-34}$ W$\cdot$s^2

$\quad = 4.13_5 \cdot 10^{-15}$ eV$\cdot$s   Planck's constant, where

$e = 1.60_2 \cdot 10^{-19}$ A$\cdot$s   elementary electronic charge

Notes:

1. For calculations the same units of length must be used in all terms.

2. The definitions (b) to (e) are mostly used in physics where it is necessary to give the relationship between the spectral property of matter and the light energy or a quantity that is directly proportional to energy.

Conversion values of quantities used in the specification of wave

Wavelength	Frequency	Photon	Wave number
λ	ν	E_{ph}	$\bar{\nu}$
nm	$10^{14}s^{-1}$	eV	$10^3 cm^{-1}$
250	11.99	4.96	40.0
300	9.99	4.13	33.3
350	8.57	3.54	28.6
400	7.49	3.10	25.0
450	6.66	2.75	22.2
500	6.00	2.48	20.0
550	5.45	2.25	18.2
600	5.00	2.07	16.7
650	4.61	1.91	15.4
700	4.28	1.77	14.3
750	4.00	1.65	13.3
800	3.75	1.55	12.5
850	3.53	1.46	11.8
900	3.33	1.38	11.1
950	3.16	1.31	10.5
1000	3.00	1.24	10.0
1050	2.86	1.19	9.52
1100	2.73	1.13	9.09

2. Quantities, Units and Symbols

2.1 Alphabetical Order of Symbols and Units

Symbols for Quantities

Letters of the English Alphabet

a	Amplitude of light waves
a	Length
a	Linear absorption coefficient
A	Absorbance
b	Half of the long edge of a rectangular aperture area
B	Radiance
c	Velocity of light in vacuo (effectively also in air)
c	Concentration of mass in a solution
C	Proportionality constant
d	Distance perpendicular to the optical axis in the microscope
DF	Depth of focus
E	Irradiance
E_{ph}	Photoelectric energy
e	Elementary electronic charge (in $A \cdot s$)
f	Focal length
f	Calculation factor

F	Area (German: Fläche)
FN	Field-of-view number of the ocular
g	Grating constant of a monochromator grating
G	Measuring value, originally applied to 'galvanometer reading'
G	Gain in the photomultiplier tube
h	Planck's constant
I	Intensity of light; this term can be applied to all radiometric and photo-metric quantities
IF	Interaction factor; this measures a property of matter interacting with light
k	Absorption coefficient
k	Order of diffraction in a diffraction spectrum or an integer
k	Boltzmann's constant
K	Luminous efficacy
I	Length (distance) parallel to the optic axis in the microscope or in another specified direction
LC	Linear conductivity
m	Total magnification, that is, scale factor or angular magnification of an image produced by several lenses
m	Mass
M	Numbers of grooves in a diffraction grating
MN	Magnification number of the objective (subscript ob) or the ocular (subscript oc); portion of the magnification with which the objective or the ocular contributes to the total magnification (m)
n	Refractive index of the stage object or a feature in the stage object
N	Refractive index of material other than the stage object
N	Any number
NA	Numerical aperture (unless otherwise specified the NA refers to the aperture angle that is effective for the transfer of light energy at practical work)
OF	Optical flux
OP	Optical path
OPD	Optical-path difference
p	Fraction of a field occupied by the stage object
p	Ratio of the measuring value for the specimen to the measuring value for the rebference material
P	Fraction of light lost by reflection at the faces of a plate
q	Ratio of the incremental measuring value to the measuring value for the reference material
Q	Quotient of two specified quantities
Q	Quantity of radiant energy
r	Radius of an area perpendicular to the optical axis of the microscope
R	Reflectance of the stage object (unless specified specular reflectance in air at normal incidence is involved)
S	Blackening of photoemulsion (German: Schwärzung)
S	Sensitivity, quotient of the output quantity to input quantity or to a quantity related to the input quantity

t	Thickness of material situated in the light train (along the direction of the axis of the light beam)
T	Transmittance of the stage object
T_c	Colour temperature
TF	Tube factor
u	Aperture angle
v	Obliquity of illumination
$V(\lambda)$	Luminous efficiency
w	Field angle
x	General symbol for an unspecified measurable quantity

Greek Letters

α	Angle between two specified directions
α	Specific refraction increment
β	Absorptance
β	Prism angle of a monochromator prism
γ	Angle between two specified directions
ε	Absorptivity
κ	Absorption index
λ	Wavelength
ν	Frequency
$\bar{\nu}$	Wave number
φ	Phase angle of light waves
Φ	Radiant flux
ρ	Reflectance of glass faces situated in the light train
σ	Standard deviation
τ	Transmittance of optical elements situated in the light train, except the stage object.
ω	Angular frequency
ω	Solid angle

Symbols for Units

A	Ampère
g	Gram mass
lm	Lumen
lx	Lux
m	Metre
mol	Mole $=$ g times 'molecular-weight'
rad	Radian
s	Second (of time)
sr	Steradian
V	Volt
W	Watt
$^\circ$	Degree angle $[\triangleq(\pi/180)\text{rad}]$
$^\circ$C	Degree Celsius
K	(Degree) Kelvin $(=273.15+\text{Degree Celsius})$

Prefixes for Decimal Multiples of Units

c   centi 10^{-2}
m   milli 10^{-3}
μ   micro 10^{-6}
n   nano 10^{-9}

2.2 Modifying Signs

The use of modifying signs is mostly restricted to those signs that are standard-ised; sometimes a non-standardised sign is modified for clarity.

Subscripts

The subscripts refer to:
    a) module of the equipment (e.g. ob; objective),
    b) object of the measurement (e.g. sp; specimen),
    c) a certain value within a range of values including maximum (max) and minimum (min) (e.g. o; central value),
    d) a certain state or condition (e.g. p; parallel, r; reflected).

For special meaning of the subscripts v (visual) and λ, see Appendix 2.5.

Superscripts to the Right of the Symbol

Asterisks indicate a quantity the value of which is not correct. Inverted commas indicate a quantity in a sequence of quantities of the same kind. The superscript to the left of R refers to the medium in which reflectance is measured (e.g. oil means standard immersion oil).

Parentheses

(λ) indicates spectral distribution; if the quantity is related to a special wave-length, it is replaced by that wavelength (e.g. 546) in nm.

Prefixes

Δ a range of the quantity between specified limits e.g. $\Delta\lambda$, or an error in a quantity or factor (difference between the false and the correct value of a quantity or factor, e.g. ΔR).

Notes:
1. The necessary symbols are explained for each equation.

2. The quantities, units and symbols adopted in this book have, where possible and convenient, been made consistant with those recommended by international agreement (German Standards DIN 5031, 5032; International Lighting Vocabulary, The Symbols Committee of the Royal Society, 1975). Some symbols have been modified with the aspect of most convenient adaption to the topic of this book.

2.3 Mathematical Symbols

$=$	equal to
$\equiv$	identically equal to
$\hat{=}$	corresponds to
$\approx$	approximately equal to
$\rightarrow$	approaches
$\propto$	proportional to
∞	infinity
$\leq$	smaller than or equal to
$\geq$	larger than or equal to
$+$	plus
$-$	minus
$\pm$	plus or minus
$a \cdot b$	a multiplied by b
$a/b, \dfrac{a}{b}, a \cdot b^{-1}$	a divided by b
(a^n)	a raised to power n
$\sqrt{a}$	square root of a
$\bar{a}$	mean value of a
$f(x) = y$	function of x
$\lim\limits_{x \to a} f(x)$	limit to which $f(x)$ tends as x approaches a
Δx	finite increment of x or finite difference $x_1 - x_2$ including an error in x
dx	differential of x
dy/dx	differential coefficient of y with respect to x
$\displaystyle\sum_{x=a}^{b} y \cdot \Delta x$	sum of the products $(y \cdot \Delta x)$ from $x=a$ to $x=b$, where $(b-a)$ is divided into equal intervals (Δx) and $y = f(x)$
$\displaystyle\int_{a}^{b} y \cdot dx$	definite integral of y from $x=a$ to $x=b$
$e^x = \exp(x)$	exponential of x
e	base of natural or Napierian logarithms $(=2{,}7183)$
$\lg x$	logarithm to the base 10 of x $(\log_{10} x)$ (common logarithm of x)
$\ln x$	natural logarithm of x $(\log_e x)$
$\sin x$	sine of x
$\cos x$	cosine of x
$\tan x$	tangent of x

2.4 Quantities in the Interaction of Light and Matter (Alphabetical Order)

Quantity	Symbol	Definition	Explanation and notes
Absorbance (decadic)	A	$\lg(1/T_i)$ or $I_t/I_{ent} = 10^{-A}$	T_i (fraction) I_{ent} intensity of light entering the substance I_t intensity of light after a path of length t
Absorptance	β	$1 - T_i$	T_i (fraction)
Absorptivity	ε	$[A/(c \cdot t)] \, \text{cm}^2 \cdot \text{g}^{-1}$ or $\text{cm}^2 \cdot \text{m mol}^{-1}$	c concentration (in $\text{g} \cdot \text{cm}^{-3}$ or $\text{m mol} \cdot \text{cm}^{-3}$) t (in cm) If necessary a distinction is made between normal absorptivity related to g mass and molar absorptivity related to m $\cdot$ mol, where mol is molecular weight unit
Absorption coefficient	k	$I_t/I_{ent} = e^{-\frac{4\pi k t}{\lambda}}$	I_{ent} intensity of light entering the substance I_t intensity after a path of length t
Absorption index	κ	$I_t/I_{ent} = e^{-\frac{4\pi \kappa t}{\lambda_s}}$	The wavelength (λ) is in air (vacuum) in the first and inside the substance (λ_s) in the second case, where $\lambda/\lambda_s = n$ or $k = n \cdot \kappa$
Blackening (German: Schwärzung)	S	$\lg(1/T_i)$	This name is applied to absorbance of an exposed and developed photographic emulsion
Diffuse reflectance	R_{diff}	I_{diff}/I_i	I_i intensity of incident light I_{diff} intensity of diffusely reflected light, i.e. of light reflected in the whole room above the substance. This room has solid angle 2π sr. The amount of R_{diff} depends on the thickness of the reflecting material and on the geometry of the incident beam, so the thickness and geometry must be specified. Unless the reflected light is accepted by an integrating sphere, also the geometry of the reflected beam accepted by the optical system must be specified
Linear absorption coefficient	a	$(A/t) \, \text{cm}^{-1}$ or $I_t/I_{ent} = 10^{-a \cdot t}$	Absorbance per unit thickness of the substance or unit path length of light (t) (in Lambert's law of absorption) t usually in cm (Not to be confused with the absorption coefficient)
Optical path	OP	$n \cdot t$	Product of (mechanical) thickness and refractive index of a substance
Optical-path difference	OPD	$(n_1 t_1 - n_2 t_2)$	Difference of optical paths in the media 1 and 2
Optical phase	φ	$n \cdot t(360°/\lambda)$ or $n \cdot t(2\pi \, \text{rad}/\lambda)$	t and λ must be given in the same unit
Optical-phase difference	$\Delta\varphi$	$(n_1 t_1 - n_2 t_2) \, 360°/\lambda$ or $(n_1 t_1 - n_2 t_2) \dfrac{2\pi \, \text{rad}}{\lambda}$	

Quantity	Symbol	Definition	Explanation and notes
Reflectance	R (ρ)	I_r/I_i	I_i intensity of incident light I_r intensity of light reflected into a specified direction Unless otherwise specified R is referring to specular reflection at vertical incidence (for distinction of R and ρ see App. 2.1)
Reflectivity	R_∞	$(I_{diff}/I_i)_\infty$	Diffuse reflectance of material of infinite thickness
Refractive index	n N	c/c_s	c velocity of light in vacuo c_s velocity of light in the substance If an absorbing substance is concerned, c_s is the phase velocity, i.e. the velocity with which a certain phase within a group of waves is moving in and together with the group (For distinction between n and N see App. 2.1)
Transmittance (total)	T	I_e/I_i	I_e intensity of light leaving the substance I_i intensity of light incident on the substance Total transmittance is referred to a substance of the same kind as that to which internal transmittance is referred, but taking into account loss of light by reflection at the entrance and exit face
Transmittance (internal)	T_i (τ_i)	I_t/I_{ent}	I_{ent} intensity of light entering the substance I_t intensity of light after a path of length t in the substance The path may be identically equal to the thickness of the specimen so that I_t is the intensity just reaching the exit surface This term is, unless otherwise specified, used for a clear substance, i.e. a substance causing no diffusion, and excluding loss by reflection at faces. The subscript i and the adjective internal are usually omitted in practice. (For distinction between T and τ see App. 2.1)

Notes:

1. If the number e is used as the base of a logarithmic quantity, the adjective 'natural' is applied to this quantity and its symbol is modified by the subscript n; such as, for example, natural absorbance; $A_n = \ln(1/T)$.

2. The linear natural absorption coefficient a_n is, in physics, usually called 'absorption constant': the reciprocal of this quantity is called 'mean penetrating depth' $(1/a_n = t)$; this is the path length in the substance for which the intensity of the entering light is reduced to a fraction of $\exp(-1) = 0.37$, according to the expression $I_t/I_{ent} = \exp(-a_n \cdot t) = \exp(-1)$.

3. The mean penetrating depth is used for distinguishing between 'weak' and 'strong' absorption. For weak absorption the mean penetrating depth must be greater than one wavelength in vacuo; thus $1/a_n > \lambda$ or $k < 0.08$, for strong absorption is must be smaller, or $k > 0.08$.

4. Transmittance or absorbance cannot be used for the distinction between 'weak' and 'strong' absorption because each substance can be made to have strong transmittance (or weak absorbance) if it is made sufficiently thin.

5. Sometimes reflectivity is used as a synonym for reflectance, but the name reflectivity should be restricted to the definition given here.

6. Unfortunately, in paper industries the name blackening is defined in another way than in the previous list; it is applied to the ratio of the diffuse reflectance of a sheet of given thickness to the diffuse reflectance of a sheet of infinite thickness (reflectivity).

7. In practice, often no distinction is made between internal and total transmittance and the subscript i is omitted for internal transmittance. This is tolerated if there is sufficient evidence for the character of transmittance. Such evidence exists, for example, if the transmittance of a substance is measured by comparison with a substance (reference material) of the same nature; in this case it is always internal transmittance that is involved.

Mathematical Relations between Quantities Given in the Above List

There are mutual mathematical relations between those quantities in the list applying to clear (i.e. non-diffusing) substances

Transmittance

$$T_i = 10^{-A} = 10^{-a \cdot t} = 10^{-\varepsilon \cdot c \cdot t},$$

$$T_i = e^{-A_n} = e^{-a_n \cdot t} = e^{-\varepsilon_n \cdot c \cdot t} = e^{-\frac{4 \cdot \pi \cdot k \cdot t}{\lambda}} = e^{-\frac{4 \cdot \pi \cdot \kappa \cdot n \cdot t}{\lambda}}.$$

These relations are called either Lambert's law or Lambert-Beer's law

Absorbance

$$A = a \cdot t = \varepsilon \cdot c \cdot t = \lg \frac{1}{T_i} = -\lg T_i,$$

$$A_n = a_n \cdot t = \varepsilon_n \cdot c \cdot t = \frac{4 \pi k \cdot t}{\lambda} = \frac{4 \pi \kappa \cdot n \cdot t}{\lambda} = \ln \frac{1}{T_i} = -\ln T_i,$$

$$\frac{A_n}{A} = \frac{a_n}{a} = \frac{\varepsilon_n}{\varepsilon} = 2.30_3.$$

Reflectance for normal incidence

$$^x R = \frac{(n-N)^2 + k^2}{(n+N)^2 + k^2} \quad \text{and,} \quad \text{for } k=0, \quad ^x R = \left(\frac{n-N}{n+N} \right)^2.$$

The asterisk indicates the medium against which the reflection takes place and which has refractive index N [see Eq. (9.13)].

Relation between total transmittance (T), internal transmittance (T_i), reflectance in air with normal light incidence at the entrance and exit face (R), and refractive index (n) of a clear plate with plane-parallel faces;

$$T = \frac{(1-R)^2}{1 - R^2 \cdot T_i^2} \cdot T_i.$$

The loss of light due to reflections at the entrance and exit face is approximately expressed by the 'reflection factor' $(2 \cdot n)/(n^2 - 1)$ so that the relation between total and internal transmittance is expressed by

$$T \approx T_i \frac{2n}{n^2 - 1}.$$

Note: No attempt has been made to clear up the confusion about the correct name of the formulas expressing *R*. In the literature the following names will be found: Beer's formula, Cauchy's formula, Fresnel's formula and Young's formula.

2.5 Radiometric and Photometric Quantities (Alphabetical Order)

Notes for table on opposite page:

1. In the units of radiometric quantities the areas are explicitly expressed and can be given in any unit; in units of photometric quantities in which areas are implicit, the unit for area is m^2, thence for the transformation between radiometric and photometric quantities the conversion of area units must be taken into account, e.g. $10^4 \, cm^2 = m^2$ for the quantities listed above.

2. In the field of photography the common logarithm of the quantity of illumination is called exposure.

3. The letter v (visual) is used as a subscript when required to distinguish photometric quantities.

4. Sometimes the luminance of the luminous surface of a lamp is expressed in terms of colour temperature (with the unit K). This is the temperature of a black body (full radiator) which emits radiation of the same chromaticity as the radiation from the source studied. The chromaticity is expressed by coordinates in the CIE (Commission Internationale de l'Eclairage) Chromaticity Diagram and in this diagram the chromaticities for a set of colour temperatures are plotted (see German Standard DIN 5033). But it must be mentioned:

a) For a given colour temperature, the relative spectral distribution of spectral concentration of radiance of the black body and the lamp may be different (see Fig. 4.2a) so that the colour temperature gives no information about the spectral distribution of radiometric or photometric quantities of a luminous surface other than the black body.

b) The colour temperature of a tungsten filament is always higher than its true temperature (Forsyth and Worthing, 1925).

Transformation between Radiometric and Photometric Quantities

Basic expression for the photometric evaluation of radiation

$$I_v = K_{max} \cdot \int_{380\,nm}^{780\,nm} I_\lambda(\lambda) \cdot V(\lambda) \cdot d\lambda \tag{A.1}$$

I any of the radiometric quantities listed above [for subscript λ and paranthese λ see note (2)]

I_v the corresponding photometric quantity (v visual)

$V(\lambda)$ spectral luminous efficiency of the human eye (fraction) [see note (3)]

K_{max} maximum luminous efficacy (in $lm \cdot W^{-1}$) [see note (4)]

In practice the integral is replaced by a summation with the limits of the spectral range restricted and the range $d\lambda$ widened to a finite size $\Delta\lambda$, being

Optical Radiation Physics			Definition	Illumination Engineering		
Symbol	Unit	Radiometric quantities		Photometric quantities	Symbol	Unit
$\dfrac{\Phi}{F} \equiv E$	$W \cdot cm^{-2}$	Irradiance	Radiant flux per unit area of receptor	Illumination or illuminance	$\dfrac{\Phi_v}{F} \equiv E_v$	$lm \cdot m^{-2} \equiv lx$
$E \cdot time$	$W \cdot s \cdot cm^{-2}$	Irradiation	Amount of irradiance within a certain time	Quantity of illumination	$E_v \cdot time$	$lm \cdot s \cdot m^{-2} = lx \cdot s$
B	$W \cdot cm^{-2} sr^{-1}$	Radiance	Radiant flux per unit area (of emitter or receptor) and per unit solid angle	Luminance or photometric brightness	B_v	$lm \cdot m^{-2} \cdot sr^{-1} = lx \cdot sr^{-1}$
Φ/F	$W \cdot cm^{-2}$	Radiant emittance or radiant excitance	Radiant flux per unit area of emitter	Luminous excitance or luminous emittance	Φ_v/F	$lm \cdot m^{-2} = phot$
Φ	W	Radiant flux	Power (i.e. energy per unit time) emitted, transferred, or received as light (radiation)	Luminous flux	Φ_v	lm
Q	$W \cdot s$	Quantity of radiant energy	Energy emitted, transferred or received as light (radiation)	Quantity of light	Q_v	$lm \cdot s$

not less than 5 nm. The corresponding expression is

$$I_v = K_{max} \cdot \sum_{\lambda=400\,nm}^{700\,nm} I_\lambda(\lambda) \cdot V(\lambda) \cdot \Delta\lambda.$$

(A.2)

Notes:

1. Equation (A.1) is true for specified photometric conditions. Specification is found in standard papers, e.g. DIN 5032.

2. The subscript λ indicates the 'spectral concentration' or spectral density of a radiometric quantity; thus $I_\lambda = I/\Delta\lambda$, the unit being that of the quantity in question times nm^{-1}.

The parenthesis λ indicates the 'spectral distribution,' i.e. a value for a particular wavelength. For example, $I_\lambda(\lambda)$ indicates spectral distribution of the spectral concentration of I.

The term 'spectral' by itself refers to the spectral distribution as in spectral transmittance $[T(\lambda)]$ for example.

3. The maximum value of $V(\lambda)$ is 1; it occurs at 555 nm $\equiv \lambda_{max}$ for vision with bright-adapted eye (photopic vision) or at 507 nm $\equiv \lambda_{max}$ for vision with dark-adapted eye (scotopic vision). The spectral distribution of luminous efficiency is shown in spectral curves (Fig. A.1). $V(\lambda)$ is based on the following measuring procedure: in a visual spectrophotometer a beam of light of wavelength λ_{max} has radiant flux $\Phi(\lambda_{max})$. Another beam of light of wavelength λ is adjusted in power until it makes the same luminous sensation in the eye of the observer; this has radiant flux $\Phi(\lambda)$. The value is then

$$V(\lambda) = \Phi(\lambda_{max})/\Phi(\lambda)$$

(A.3)

$\Phi(\lambda_{max})$ is the radiant flux for the wavelength at which the curve (not the radiant flux!) has its maximum, thus $\Phi(\lambda_{max}) < \Phi(\lambda)$. Data for $V(\lambda)$ are available in standard papers, e.g. DIN 5031.

4. The maximum luminous efficacy is the luminous flux produced by unit radiant flux and observed under specified conditions: the quantity is 673 lm·W^{-1} at 555 nm for photopic vision or 1725 lm·W^{-1} at 507 nm for scotopic vision. These quantities are calculation constants.

5. A difference is made between the terms 'efficiency' and 'efficacy.' The former is used for the ratio of terms that have no units or have the same units, so that efficiency has no unit; the latter for the quotient of terms that have different units, hence the efficacy is given in a unit.

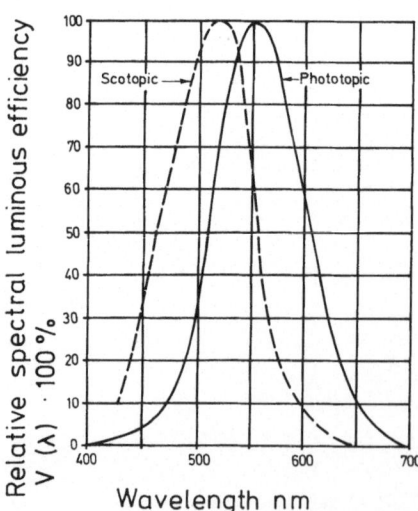

Fig. A.1. International luminous-efficiency curves for the bright- and dark-adapted human eye (photopic and scotopic vision)

3. Accumulation of Errors

In the accumulation of errors both types are involved; systematic errors are differences between the actual (false) measuring value or result and the true value or result; statistical errors are expressed as variances (σ^2). The mathematical description of the accumulation applies to both types of error, and so we can use similar formulas.

In the case of comparison measurements, there are two terms in the numerator and one in the denominator; since each has its own error, we have to use the differential calculus in order to obtain the error for the measuring result (interaction factor of the specimen),

$$IF_{sp} = \frac{G_{sp}}{G_{ref}} \cdot IF_{ref} \tag{9.4}$$

IF true interaction factor
G true measuring value
sp   specimen
ref   reference material

We take each of the quantities on the right side of Eq. (9.4) as being affected with an error. The systematic integral error of the measuring result is given by

$$\Delta(IF_{sp}) \equiv \Delta\left(\frac{G_{sp}}{G_{ref}} \cdot IF_{ref}\right) = \frac{G_{sp}}{G_{ref}} \cdot \Delta(IF_{ref}) + \frac{IF_{ref}}{G_{ref}} \cdot \Delta G_{sp} + \frac{G_{sp}}{G_{ref}^2} \cdot IF_{ref} \cdot \Delta G_{ref} \tag{A.4}$$

Δ Systematic error expressed by $\Delta(IF) = IF^* - IF$; $\Delta G = G^* - G$

The corresponding relative integral error of the measuring result is given by

$$\frac{\Delta(IF_{sp})}{IF_{sp}} = \frac{\Delta(IF_{ref})}{IF_{ref}} + \frac{\Delta G_{sp}}{G_{sp}} + \frac{\Delta G_{ref}}{G_{ref}}. \tag{A.5}$$

The statistical error of the measuring result is given by

$$\sigma_{IF_{sp}}^2 = \left(\frac{G_{sp}}{G_{ref}}\right)^2 \cdot \sigma_{IF_{ref}}^2 + \left(\frac{IF_{ref}}{G_{ref}}\right)^2 \cdot \sigma_{G_{sp}}^2 + \left(\frac{G_{sp}}{G_{ref}^2} \cdot IF_{ref}\right)^2 \cdot \sigma_{G_{ref}}^2 \tag{A.6}$$

σ^2 variance of the quantity indicated by the subscript. For calculations the mean values of the quantities must be used.

The corresponding relative statistical error is

$$\left(\frac{\sigma_{IF_{sp}}}{IF_{sp}}\right)^2 = \left(\frac{\sigma_{IF_{ref}}}{IF_{ref}}\right)^2 + \left(\frac{\sigma_{G_{sp}}}{G_{sp}}\right)^2 + \left(\frac{\sigma_{G_{ref}}}{G_{ref}}\right)^2. \tag{A.7}$$

We can state that, when the measuring result is either the product or the quotient of the terms, then the relative error is the sum of the relative errors of the terms. In the special case where only one term has an error, the others being accurate or precise, the relative error of the measuring result is the same as that of this one term.

Error formulas are listed in mathematical formula tables, the derivation of the formulas is found in the literature on mathematical error analysis. (See also Ku, 1960.)

4. Detectivity and the Effect of Statistical Spread

Definitions

Detectivity is expressed as the minimum quantity of anything that can be seen to affect the measuring value. We take the detectivity as being proportional to the reciprocal of the sensitivity, for which see Chapter 3, Section 7, and the proportionality factor as being the range of confidence of the mean value (Chap. 13, Sect. 2). In the same way we make a distinction between 'absolute detectivity' and 'minimum absolute detectivity' of a quantity; for the latter we can employ the concept of 'limit of detection'; with the best setting of the whole apparatus which is the setting for maximum sensitivity, we can specify the smallest absolute amount that can be detected, and this is the limit of detection. We can define the following in terms of a quantity x,

$$x_d = x \cdot \frac{\sigma_G \cdot f_N}{\bar{G} \cdot \sqrt{N}}, \tag{A.9}$$

$$x_L = x \cdot \frac{\sigma_{G\,max} \cdot f_N}{G_{max} \cdot \sqrt{N}} \tag{A.10}$$

x_d absolute detectivity in terms of the quantity x

x_L limit of detection in terms of the quantity x

x the quantity to be detected

$\bar{G}$ the mean of a set of measuring values

$\bar{G}_{max}$ the largest mean measuring value obtainable from the photometer (at maximum sensitivity of photometric equipment)

$\sigma_N \cdot f_N / \sqrt{N}$ the range of confidence of the mean value, where σ is the standard deviation of the set of measuring values (the subscript indicates the set) in the unit of the measuring value

N number of measuring values per set

f_N calculation factor depending on N and on the statistical certainty for the range of confidence (the certainty is mostly taken as being 99,7%) (for values of f_N see Table 13.1)

Note: For the definition of detectivity according to Eqs. (A.9) and (A.10) the values of $\bar{G}$ must be strictly proportional to x.

Minimum Detectable Radiant Flux

Let us now consider what is involved in expressing the minimum radiant flux that can be measured with a given apparatus. The author carried out a test with a microscope photometer, the efficiency of which can be considered as good as any offered on the market today.

The following procedure was applied for the test and this can be done with any other microscope photometer:

1. Setting of the photoelectric assembly for maximum sensitivity.

2. Illumination of the photocathode through the microscope and control of the amount of radiant flux so that the mean measuring value was lying in the central part of the measuring scale.

3. Comparison of radiant flux with the measuring value and calculation of the limit of detection according to Eq. (A.10), i.e. expressing X in terms of Φ,

$$\Phi_L = \Phi \cdot \frac{\sigma_{G\,max} \cdot f_N}{\overline{G}_{max} \cdot \sqrt{N}} . \tag{A.11}$$

The determination of the amount of radiant flux is incertain in the same degree as are the values of parameters involved in the transfer of light through the microscope [Eq. (3.1)]. The parameters and estimated values of these for our test are listed in Table A.1. Putting the values into Eq. (3.1) we obtain the radiant flux as

$$\Phi = B_{\lambda} \cdot \Delta\lambda \cdot OF \cdot \tau_{os} \cdot T_{st} = 9.7 \cdot 10^{-11} \text{W}. \tag{A.12}$$

(For meaning of symbols see Table A.1)

For statistical evaluation the recording of measuring values is convenient. The mean value ($\overline{G}_{max}$), the number of values (N), and the standard deviation (σ_{Gmax}) displayed at the test are shown in Figure A.2.

Putting these quantities and the values obtained by this way as well as the value of Φ into Eq. (A.11) we obtain a limit of detection of radiant flux

$$\Phi_L = 2.95 \cdot 10^{-12} \text{W}.$$

Note: As a by-product of this experiment it was found that the largest part of the integral instrument noise must be attributed to electric modules other than the photosensor.

Minimum Detectable Amount of Mass of a Component in a Biological Specimen

When the limit of detection is expressed in terms of a 'derived measuring result' that has a more complex relationship to the measuring values than the radiant flux has, we must use special definitions. In the following we determine, as an example, the minimum detectable amount of mass of a component in a biological specimen which is stained with an absorbing dye. For this purpose we have to make three assumptions:

Table A.1. Optical parameters of a microscope photometer used in testing the limit of detection of radiant flux

Subject	Parameter
Tungsten-filament lamp at colour temperature 2 800 K	Spectral concentration of radiance at $\lambda = 550$ nm (taken from Fig. 4.2 a) $$B_\lambda(550) = 10^{-2}\, \frac{W}{cm^2 \cdot sr \cdot nm}$$
Narrow-band interference filter	Spectral bandwidth $\Delta\lambda = 10$ nm
Factors affecting the optical flux: Stage-object field; radius $r = 16\,\mu m = 1.6 \cdot 10^{-3}$ cm Effective NA of 0.1	Optical flux calculated with the help of Eq. (3.20) $$OF = (r \cdot \pi \cdot NA)^2 = 2.5 \cdot 10^{-7}\, cm^2 \cdot sr$$
Factors affecting total transmittance and number of elements involved 24 coated glass surfaces (transmittance) 0.78 1 mirror (reflectance) 0.90 1 neutral filter (transmittance) 0.30 1 interference filter (transmittance) 0.2	Transmittance of whole optical system $\tau_{Os} = 4.2 \cdot 10^{-2}$
Neutral filter used as the stage object	Transmittance of stage object $T_{st} \equiv IF_{st} = 9.2 \cdot 10^{-2}$

Note: The wavelength of 550 nm is that for maximum spectral sensitivity of the equipment.

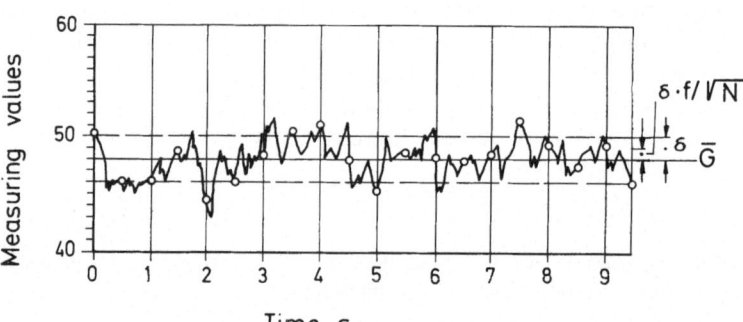

Fig. A.2. Recorded measuring values of a microscope photometer showing fluctuation due to instrumental noise. The test was carried out by the author with the equipment set for maximum sensitivity in order to determine the limit of detection of radiant flux traversing the microscope. *Small circles:* $N = 20$ measuring values taken for statistical calculations. The parameters are defined in Chapter 13, Section 2.

$\bar{G}_{20} \equiv \bar{G}_{max} = 48$ scale intervals; $\sigma_{20} \equiv \sigma_{G_{max}} = \pm 1.89$ scale intervals; $\sigma_{G_{max}} \cdot \dfrac{f_{20}}{\sqrt{20}} = \pm 1.46$ scale intervals where $f_{20} \cdot \sqrt{20} = 0.77$ for the statistical certainty of 99.7 % (see Table 13.1)

1. The molar absorptivity (ε) is, for our material, $2.5 \cdot 10^4$ cm$^2 \cdot$m mol^{-1}. This is an average value that is true for many organic dyes (Bartels, 1966).

2. The greatest reliably measurable transmittance (T) is 0.95 (95%), hence the corresponding absorbance (A) is $2.223 \cdot 10^{-2}$. These values are dictated by the smallest relative difference between the transmittance of the reference material and that of the specimen, that can be discerned. Taking this difference as being 0.05 and the transmittance of the reference material to be 1, we obtain the values above according to

$$\frac{\bar{G}_{ref} - \bar{G}_{sp}}{\bar{G}_{ref}} = \frac{T_{ref} - T_{sp}}{T_{ref}} = 1 - T_{sp} = 0.05$$

where $T_{ref} = 1$, hence $T_{sp} = 0.95$

G mean measuring value
T transmittance (fraction)
sp   specimen
ref   reference

3. The minimum diameter of the photometric field in the specimen is 0.5 μm (Zimmer, 1973), thus the minimum area of the field (F_{min}) is about 0.2 μm^2.

We put the three quantities (ε, A, F_{min}) into Eq. (9.10) and obtain the minimum detectable amount of mass approximately as

$$m_{min} = \frac{A \cdot F_{min}}{\varepsilon} \approx 2 \cdot 10^{-15} \text{ m mol}. \tag{A.13}$$

If we express the mass in grams and take a substance having a molecular weight of 300 atom mass units (amu) we obtain $2 \cdot 10^{-15} \cdot 300 \cdot 10^{-3}$ g $= 6 \cdot 10^{-16}$ g.

Note: The extremely low limit of detection of mass with the microscope photometer arises from the microscopic size of the measured volume.

5. Systematic Errors Due to Increments in Measuring Values

We call this 'incremental error.' Its sources are: photosensor dark current, glare, outside light, displacement of the zero point of the measuring scale, false light in the monochromatising device (for description see Chap. 7, Sect. 2.3; Chap. 12, Sect. 3.1; Chap. 12, Sect. 3.5). There are always two components, from the specimen and the reference material, and these are of the same magnitude, except with false light. The error is mathematically expressed and illustrated in Figure A.3.

The error due to *false light in the monochromatising device* is derived as follows:

First we express the false light by Eq. (5.1) where it is given in percentage, but we express it as a fraction. Since the false light takes the regular optical

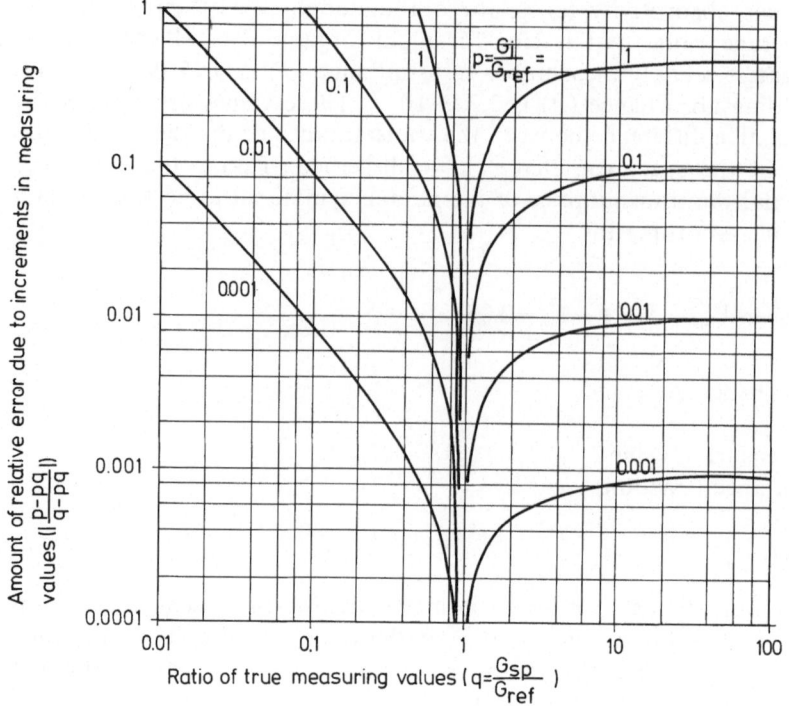

Fig. A.3. Diagram showing relative errors in the interaction factor of the specimen caused by increments in the measuring values.

Defining equation: The measuring result influenced by the increment is expressed by

$$\frac{IF_{sp}^*}{IF_{ref}^*} = \frac{G_{sp} + \Delta G_i}{G_{ref} + \Delta G_i} = \frac{G_{sp}^*}{G_{ref}^*} \tag{A.14}$$

IF interaction factor; G true measuring value, sp specimen, ref reference material; ΔG_i incremental measuring value. The asterisk indicates quantities containing the incremental error.

The ratio of true interaction factors or corresponding measuring values is expressed by the symbol q,

$$\frac{IF_{sp}}{IF_{ref}} = \frac{G_{sp}}{G_{ref}} = q \tag{9.1}$$

IF true interaction factor.

The ratio of the incremental measuring value and the true measuring value for the reference material is expressed by the symbol p,

$$\frac{\Delta IF_i}{IF_{ref}} = \frac{\Delta G_i}{G_{ref}} = p \tag{A.15}$$

ΔIF_i increment in the measuring value expressed in terms of interaction factor of the reference material.

The relative error is expressed by the difference of Eqs. (A.14) and (9.1) related to Eq. (9.1), hence by

$$\frac{\dfrac{G_{sp}^*}{G_{ref}^*} - \dfrac{G_{sp}}{G_{ref}}}{\dfrac{G_{sp}}{G_{ref}}} = \frac{\Delta G_i(G_{ref} - G_{sp})}{G_{sp}(G_{ref} + \Delta G_i)} = \frac{p - pq}{q - pq}. \tag{A.16}$$

train through the microscope the measuring value for false light (i.e. the increment) is expressed by

$$\Delta G(\lambda)_{i\,sp} \propto \widetilde{IF}_{sp} \cdot \frac{I_f}{I(\lambda)}, \tag{A.17}$$

$$\Delta G(\lambda)_{i\,ref} \propto \widetilde{IF}_{ref} \cdot \frac{I_f}{I(\lambda)} \tag{A.18}$$

and the measuring value for the stage object influenced by the false light is expressed by

$$\frac{G^*(\lambda)_{sp}}{G^*(\lambda)_{ref}} = \frac{G(\lambda)_{sp} + \Delta G(\lambda)_{i\,sp}}{G(\lambda)_{ref} + \Delta G(\lambda)_{i\,ref}} \tag{A.19}$$

G measuring value
sp   specimen
ref   reference material
ΔG_i increment
λ central wavelength of the pass band for which the monochromatising device is set
IF   interaction factor
I_f absolute intensity of false light (in any radiometric or photometric unit)
$I(\lambda)$ intensity of light having wavelengths within the pass band (in the same unit as I_f)
$I_f/I(\lambda)$ false-light portion (fraction)

The asterisk indicates that the measuring value contains the increment. The tilde indicates that the value of the interaction factor is the average value within the spectral range over which the false light is distributed.

◄   This is the general equation for the graphs in the diagram. The graphs are given for the following parameter values: $p=1$, 0.1, 0.01 and 0.001 as a function of abscissa values q in the range of $0.01-0-100$.

     The curves for values of $q<1$ are true for the condition $G_{sp}<G_{ref}$, those for values of $q>1$ for the condition $G_{sp}>G_{ref}$.

Notes:

1. In transmitted-light measurements, including those of interference, only the condition $G_{sp}<G_{ref}$ obtains. In reflected light either condition may obtain.

2. The diagram shows that (a) for a given value of q the amount of relative error increases with increasing p, that (b) for a constant value of p the amount of error decreases as q approaches unity and reaches zero for $q=1$.

3. For $q<1$ the error is positive for $q>1$ it is negative.

4. The relative error increases with the increment and also as the difference $\pm\,(G_{sp}-G_{ref})$ increases. When the measuring value is small, every effort must be made to suppress incremental errors because otherwise even a small absolute error can produce a large relative one.
The symbol G_i in the diagram means ΔG_i

For the 'non-sensitive' range (Chap. 12, Sect. 3.4) both values of $G(\lambda)_{sp}$ and $G(\lambda)_{ref}$ are zero so that, for this range, the false measuring result is expressed by

$$\frac{G^*(\lambda)_{sp}}{G^*(\lambda)_{ref}} = \frac{\Delta G(\lambda)_{i\,sp}}{\Delta G(\lambda)_{i\,ref}} = \frac{IF_{sp}}{IF_{ref}}. \tag{A.20}$$

Equation (A.20) shows that the measuring result is the same as the ratio of the average interaction factor of the specimen and of the reference material for the spectral range over which the false light is distributed.

6. Experimental Determination of Factors Involved in the Spectral Sensitivity of Measuring Equipment

The quantities influencing the magnitude and spectral distribution of measuring values are studied in Chapters 3 and 7, and here we give qualitatively the results of experiments showing the influence of the lamp, the photosensor and the objective. These results are derived from spectral curves shown in Figure A.4.

Fig. A.4a and b. Relative spectral sensitivity of a microscope photometer for specified photosensors ▶ (a) with a tungsten-filament lamp; (b) with a xenon discharge-arc lamp.

Conditions of experiment carried out by the author: *Microscope photometer:* equipped with UV-transmitting optics. *Objective:* Ultrafluar 10/0.2. *Condenser:* Objective used as condenser, Ultrafluar 10/0.2 with pinhole diaphragm inserted in the aperture area to reduce the numerical aperture of the condenser to 0.06.

Monochromators:

Monochromator	Width of exit (0.25 mm) adjusted to the effective numerical aperture of 0.06 gives bandwidth of (nm)
Quartz prism (M 4 Q III) for 240–600 nm used with blocking filters	1 at $\lambda = 250$ nm 18 at $\lambda = 600$ nm
Glass prism (M 4 G II) for 600–1100 nm	5.5 at $\lambda = 600$ nm 30 at $\lambda = 1100$ nm

Stage-object field: A circular photometer diaphragm with diameter of 2.4 mm gave the diameter of the (empty) stage object field as of 0.15 mm. *Construction of the spectral curves:* The curves were constructed from sets of measuring values read off in spectral intervals of 5 nm; where each measuring value was converted to that for a bandwidth of 1 nm in order to prevent the distortion of curves due to varying pass-band width in the monochromator. *Measuring procedure:* The measurement was started with the tungsten lamp and the photomultiplier tube of type RCA 4463. The photometric response was adjusted so that at the wavelength of maximum response the measuring value was 100. This value [indicated by x in (a)] was taken as reference for all subsequent sets of measurements. For each of the following sets of measuring values only the photomultiplier or the lamp was exchanged, the adjustment of optical equipment being left unchanged. The type of photomultiplier and lamps used for the test is indicated in the diagrams.

Note: The supply voltage for the photomultiplier and the external amplification of the photocurrent was individually adjusted for each set of measuring values so that the relative standard deviation of measuring values due to instrumental noise was about $\pm 1\%$ at the wavelength where the spectral curve for the photomultiplier in question has its maximum. (For definition of relative standard deviation see Chap. 13, Sect. 2)

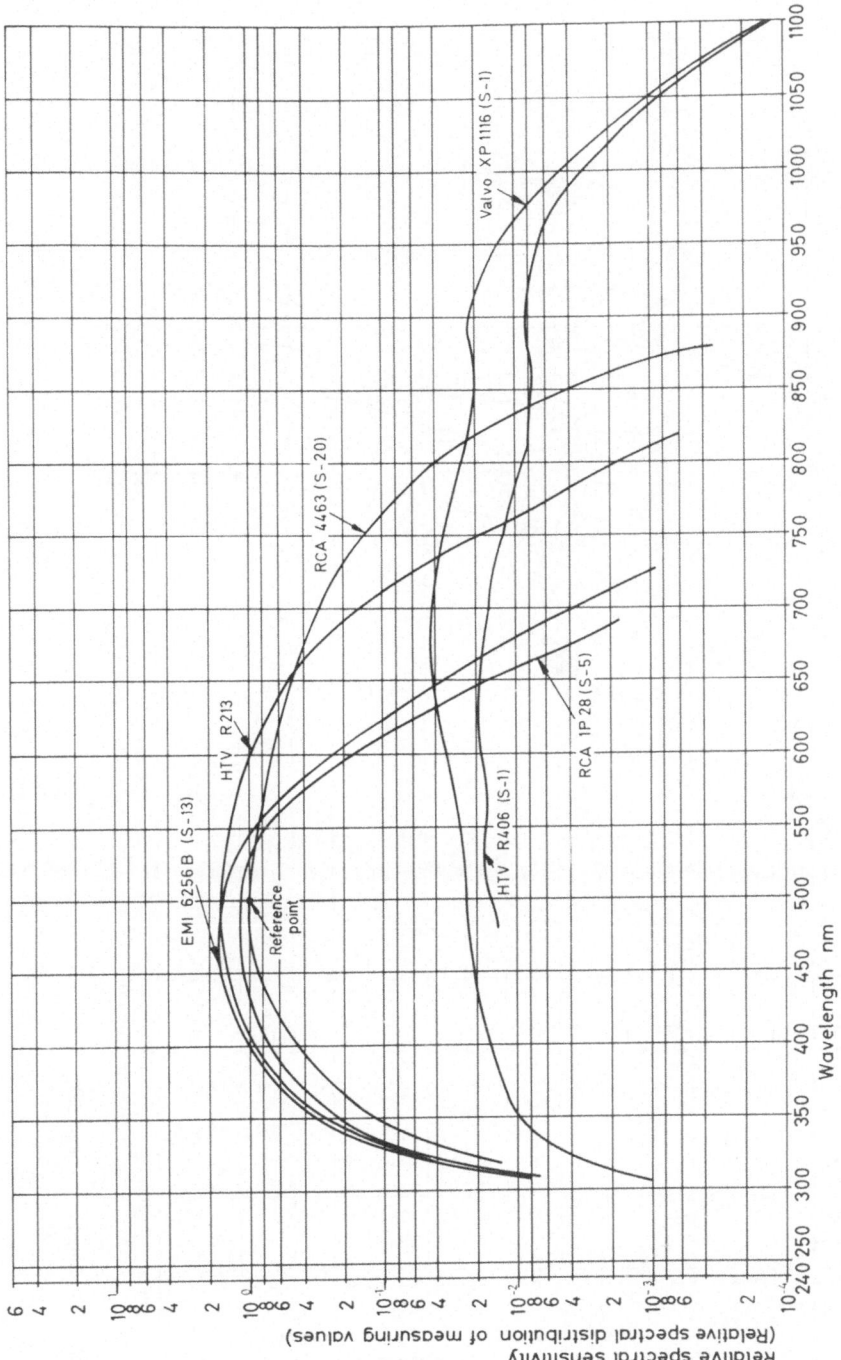

Fig. A.4a. Legend see opposite page

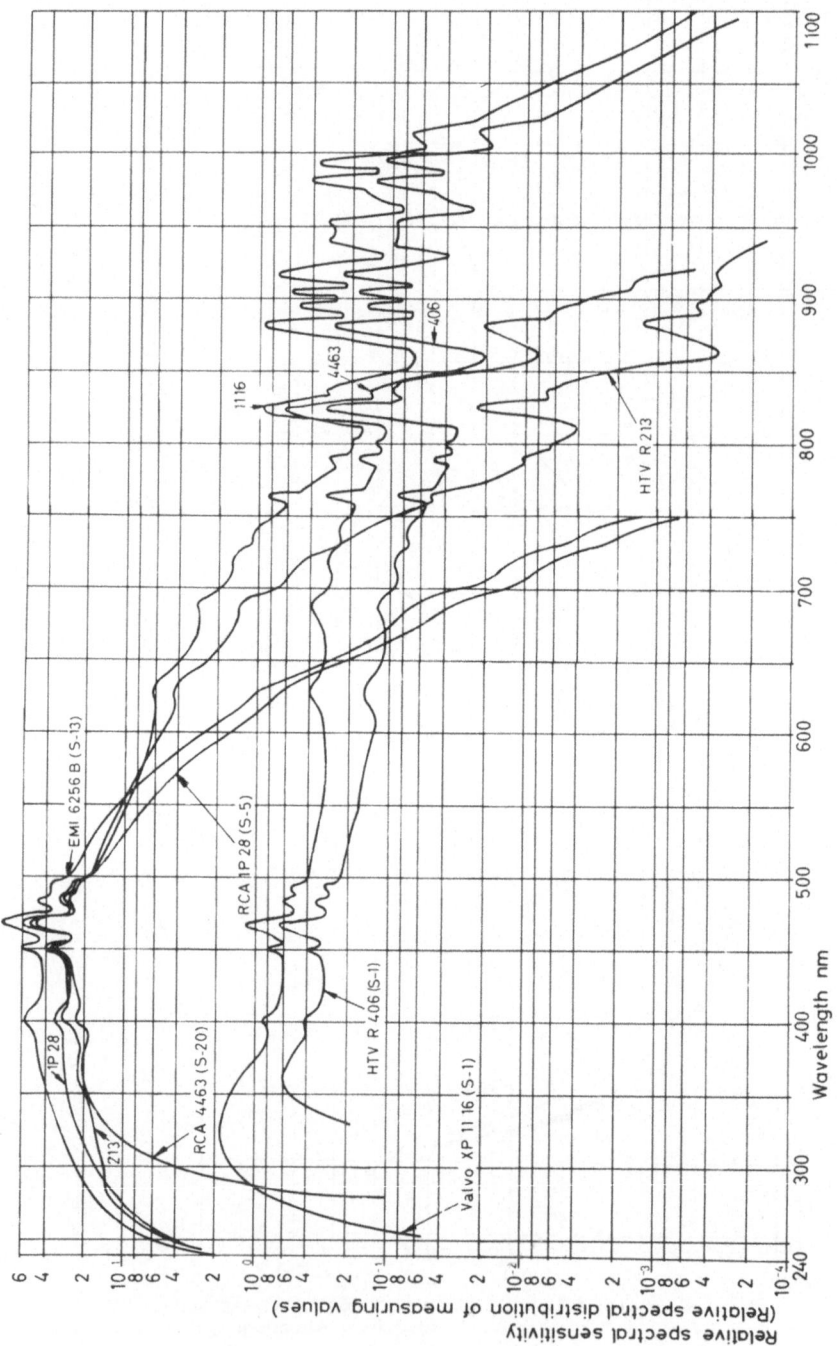

Fig. A.4b. Legend see page 220

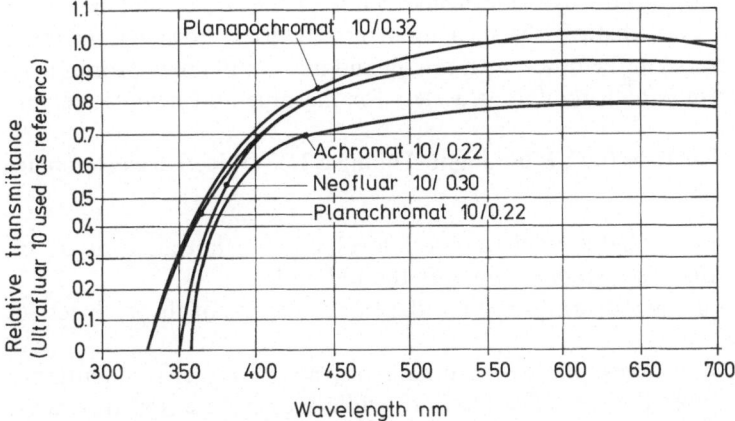

Fig. A.5. Relative spectral transmittances of various objectives. The spectral transmittances are normalised to that of an objective of type Ultrafluar.

    Conditions of experiment carried out by the author: *Microscope photometer:* Equipped with UV-transmitting optics. *Reference material:* Ultrafluar 10/0.2 (see Fig. 6.8 for absolute spectral transmittance of this objective). *Sample objectives:* Achromat 10/0.02; Planachromat 10/0.22; Neofluar 10/0.30; Planapochromat 10/0.32. *Condenser:* Ultrafluar 10/0.2 acting as condenser, with pinhole diaphragm inserted to reduce the numerical aperture to 0.18. *Monochromator:* Quartz prism (M 4 Q III) for the whole spectral range used with blocking filters. Width of monochromator exit (0.75 mm) adjusted to the effective numerical aperture of 0.18 gives the following bandwidths: 5.5 nm at $\lambda = 300$ nm; 40 nm at $\lambda = 700$ nm. *Lamp:* Xenon superpressure XBO 150. *Stage-object field:* A circular photometer diaphragm with diameter of 10 mm gave the diameter of the (empty) stage-object field as 0.63 mm. *Measuring procedure:* The spectral run was made first with the Ultrafluar and then with the sample objective in the light train at constant adjustment of apparatus.

Note: The curves do not necessarily show the typical behaviour of the types of objectives in consideration. Although the curves show the tendency towards increasing transmittance with increasing degree of correction, the transmittance of different designs of objectives having the same degree of correction can vary

Photomultiplier Tubes and Tungsten-Filament Lamp (Fig. A.4a)

1. Below about 350 nm a tungsten-filament lamp does not provide enough energy.

    2. The spectral-emission curve of the lamp is smooth; it follows that only when the stage object has peaks in the spectral curve of its interaction factor will it be necessary to reduce the bandwidth of the monochromatising device (Chap. 11, Sect. 6).

    3. It is the spectral-response characteristics of the photosensor that essentially determine the shape of the spectral curves of measuring values and the spectral limits.

Photomultiplier Tubes and Xenon (Superpressure) Lamp (Fig. A.4b)

1. The xenon lamp provides sufficient energy for the UV region.

    2. The spectral response of a given apparatus is approximately 30 times greater than with the tungsten-filament lamp; the maximum response is shifted to shorter wavelengths by 100 nm to 200 nm, depending on the type of photocathode.

3. The xenon lamp produces some feeble peaks in the violet and near UV region, but these do not disturb the measurements. It does, however, produce some pronounced peaks in the blue-green region and many still more pronounced peaks beyond 750 nm, which renders it useless for the blue and near IR region.

Transmittance of Objectives for Visible Light, Taking a UV Objective as Reference (Fig. A.5)

1. Objectives for visible light reach a good level of transmittance at about 430 nm, and this value rises slightly towards the red end.

2. In the range from 400 down to 350 nm the transmittance of these objectives goes down to zero.

3. Among these objectives, planapochromats may have a larger transmittance than objectives of lower degrees of correction, while they have a spectral range slightly extended towards shorter wavelengths.

7. Correlation Between Photometric Values and Phase Differences in Double-Beam Interference Microscopy

Here we consider the relation between the measured interference intensity and the phase difference of light waves traversing the specimen and reference material. We assume both of the latter to be perfectly transparent, so that the two traversing light waves have equal amplitudes. With polarised light both the beam splitter and the beam combiner (Fig. 10.1) have their vibration directions at an angle of 45° relative to that of the polariser. Equation (10.4) shows that the same formula applies to any kind of interference system, provided that the phase difference in question is smaller than π rad (180°).

There are two steps:

1. At given setting of the measuring equipment, we place the reference material in the measuring beam (Chap. 10, Sect. 2.2) and adjust the optical-path length in the auxiliary beam, so that a maximum measuring value is displayed. We make the measurement when the reference material shows maximum brightness, and hence the interference of light waves in the measuring beam and the auxiliary beam is perfectly constructive. The condition for this is expressed by

$$\Delta \varphi_{ref} = \varphi_{ref} - \varphi_a = k \cdot 2\pi \text{ rad}$$
$$= k \cdot 360° \qquad (A.21)$$

$$\Delta \varphi_{ref} = \varphi_{ref} - \varphi_a = (k + \tfrac{1}{2}) 2\pi \text{ rad}$$
$$= (k + \tfrac{1}{2}) \cdot 360° \qquad (A.22)$$

if the light is linearly polarised and polariser and analyser are parallel, or if the light is natural (non-polarised) (case 1).

if the light is linearly polarised and the polariser and analyser are crossed (case 2).

$\Delta \varphi_{ref}$ phase difference of light waves in the measuring beam and light waves in the auxiliary beam

φ_{ref} phase of light waves traversing the reference material in the measuring beam (in rad or degrees)

φ_a phase of light waves in the auxiliary beam (in rad or degrees)

k an integer including zero

2. We replace the reference material by the specimen, without changing anything else. This causes a shift in the phase of light waves in the measuring beam by

$$\pm \Delta\varphi = \pm(\varphi_{ref} - \varphi_{sp}) \tag{A.23}$$

$\Delta\varphi$ shift in phase (phase difference) ofight waves in the measuring beam. i.e. phase difference of light waves traversing the specimen and the reference material

φ phase of light waves

sp   specimen,

ref   reference material

This causes a shift in phase difference of the light waves in the measuring and auxiliary beam by the same amount. Expressing the initial phase difference (phase difference for perfectly constructive interference with the reference material in the measuring beam), by the terms on the right side of Eqs. (A.21) and (A.22) we can express the phase difference after the shift by

(in case 1)

$k \cdot 2\pi \pm \Delta\varphi.$

(in case 2)

$(k + \tfrac{1}{2}) \cdot 2\pi \pm \Delta\varphi.$

Next we use the basic relations between the brightness and phase difference in order to describe the whole photometric procedure mathematically. These relations are expressed by

(in case 1)

$$G \propto \cos^2 \frac{\Delta\varphi_0}{2}, \tag{A.24}$$

(in case 2)

$$G \propto \sin^2 \frac{\Delta\varphi_0}{2} \tag{A.25}$$

G measuring value (this evaluates the brightness)

$\Delta\varphi_0$ phase difference between light waves in the measuring beam with a stage object (reference material or specimen) in this beam and the light waves in the auxiliary beam

o   refers to a stage object in general

We apply Eqs. (A.24) and (A.25) to our particular procedure and replace the phase difference for the reference material ($\Delta\varphi_0 = \Delta\varphi_{ref}$) at perfectly constructive interference by the terms on the right side of Eq. (A.21) or (A.22) so that the corresponding measuring value is expressed by

(in case 1)   (in case 2)

$G_{\text{ref max}} \propto \cos^2(k \cdot \pi) = 1$ (A.26) $G_{\text{ref max}} \propto \sin^2(k+\tfrac{1}{2})\pi = 1$ (A.27)

$G_{\text{ref max}}$ measuring value for the reference material at perfectly constructive inter-
    ference

Further we express the measuring value for the specimen by

(in case 1)   (in case 2)

$$G_{\text{sp}} \propto \cos^2\left[(k \cdot \pi) \pm \frac{\Delta\varphi}{2}\right]$$ $$G_{\text{sp}} \propto \sin^2\left[(k+\tfrac{1}{2})\pi \pm \frac{\Delta\varphi}{2}\right]$$

                                                    (A.28)   (A.29)

$$= \cos^2\left(\pm\frac{\Delta\varphi}{2}\right),$$ $$= \cos^2\left(\pm\frac{\Delta\varphi}{2}\right).$$

Consequently the comparison of measuring values is, in both cases, expressed by

$$\frac{G_{\text{sp}}}{G_{\text{ref max}}} = \cos^2\left(\pm\frac{\Delta\varphi}{2}\right) = \cos^2\left(\pm\frac{OPD \cdot \pi\,\text{rad}}{\lambda}\right) = \cos^2\left(\pm\frac{OPD \cdot 180°}{\lambda}\right)$$ (10.4)

and shown in Figure A.6.

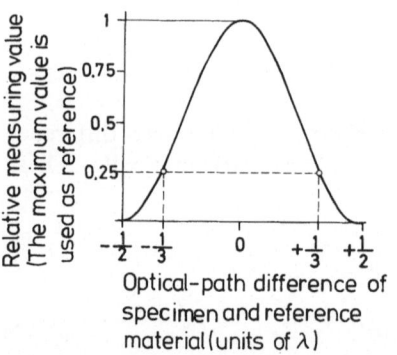

Optical–path difference of
specimen and reference
material (units of λ)

Fig. A.6. Nomogram for a double-beam interference system showing the relation between phase
difference and relative photometric measuring values referred to the maximum.
    The ratio of measuring values on the left side of Eq. (10.4) is plotted against the phase
difference $\Delta\varphi$ expressed in the terms on the right side of Eq. (10.4) within the limits of $-\lambda/$
$2 \leq OPD \leq +\lambda/2$.
    Basic expression:

$$\frac{G_{\text{sp}}}{G_{\text{ref max}}} = \cos^2\left(\pm\frac{\Delta\varphi}{2}\right) = \cos^2\left(\pm\frac{OPD \cdot \pi\,\text{rad}}{\lambda}\right) = \cos^2\left(\pm\frac{OPD \cdot 180°}{\lambda}\right)$$ (10.4)

$G_{\text{ref max}}$ measuring value (numeral) for the reference material, the interference system is adjusted
so that the perfectly constructive interference (maximum brightness in the reference material) occurs;

8. Theory of Interference-Layer Technique

The use of the interference-layer technique is described in Chapter 10, Section 6 and here we simply develop the equations. The fundamental equation (see for example, Vašiček, 1960; Pepperhoff and Ettwig, 1970) is

$$R_{min} = \left(\frac{a_1 - a_2}{1 - a_1 \cdot a_2}\right)^2 \tag{A.30}$$

R_{min} minimum reflectance due to perfectly destructive interference of light waves reflected at the layer-air surface and those reflected at the specimen-layer surface (fraction)

a_1 amplitudes of light waves after reflection at the layer-air surface

a_2 amplitudes of light waves after reflection at the specimen-layer surface

We replace the amplitudes by the square roots of the reflectances to give

$$R_{min} = \left(\frac{\sqrt{\rho_1} - \sqrt{{}^1R}}{1 - \sqrt{\rho_1 \cdot {}^1R}}\right)^2 \tag{A.31}$$

ρ_1 reflectance at the layer-air surface (fraction); $a_1 = \sqrt{\rho_1}$

${}^1R$ reflectance at the specimen-layer border face (fraction); $a_2 = \sqrt{{}^1R}$

1   layer

    In order to determine the refractive index and the absorption coefficient of the specimen we require to know the reflectance of the specimen against air and against the layer; the first we measure in the usual way and the second we derive by rearranging Eq. (A.31) for which there are the three Eqs. (A.34), (A.35) and (A.36) given in Figure A.7.

    We refer to Eqs. (9.14) and (9.15) and replace ${}^xR$ by ${}^1R$, N by N_1, and use for ${}^1R$ the terms on the right side of Eqs. (A.34) and (A.35). The following expressions then give the refractive index (n) and absorption coefficient (k)

G_{sp} measuring value for the specimen; $\Delta\varphi$ phase difference between light waves traversing the specimen and those traversing the reference material (in radians or degrees); OPD optical path difference in the specimen and the reference material (in the unit of thickness or wavelength); λ wavelength in the same unit as that of thickness. Example (indicated by the dotted straight lines): We take the measuring value for the specimen (G_{sp}) to be of 0.25 of that for the reference material ($G_{ref\,max}$). Consequently from Eq. (10.4) we get

$$\frac{G_{sp}}{G_{ref\,max}} = 0.25 = \cos^2\left(\pm\frac{\pi}{3}\right).$$

Putting the value of $\Delta\varphi/2 = \pm\pi/3$ on the right side of Equation (10.4), we obtain the value of $OPD = \pm\lambda/3$, and taking $\lambda = 546$ nm, we obtain $OPD = \pm 182$ nm

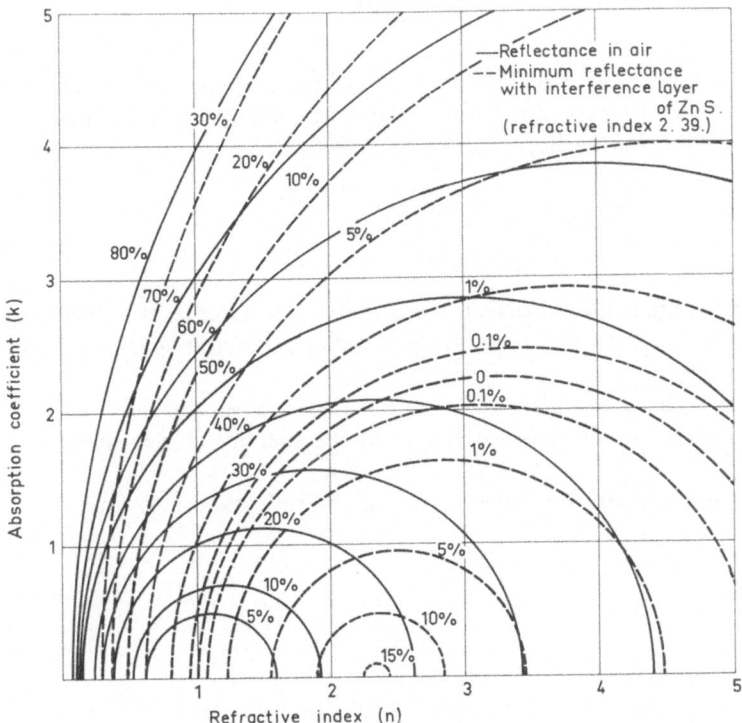

Fig. A.7. Iso-reflectance curves for the interference-layer method. In Figure 9.6 are given the iso-reflectance curves (sections of circles) for measurements in two media, air and oil. In the present figure the curves for reflection in air (*full lines*) are repeated but the second set of curves (*broken lines*) is that for the interference-layer method; they give the minimum reflectances of a ZnS interference layer on a polished surface. As in the case of Figure 9.6 the intersection of a curve for air and a curve for the layer defines n and k for the substrate material.

Note: In this case there is a zero curve, which corresponds to perfectly destructive interference; on either side of this the reflectance rises. This produces ambiguity in those points on the diagram where a given curve for air intersects two of the other curves having the same absolute value of reflectance.

Mathematical expression for the dashed curves. Coordinate for the centres of the circles;

$$n = N_1 \frac{1 + {}^1R}{1 - {}^1R}; \quad k = 0. \tag{A.32}$$

Lengths of radii;

$$r = \frac{2 N_1 \cdot \sqrt{{}^1R}}{1 - {}^1R} \tag{A.33}$$

N_1 refractive index on the layer substance (the value of 2.39 is used for the diagram; ${}^1R$ reflectance of the substrate-layer border face (fraction)

On the other hand, after rearranging Eq. (A.31), we can express ${}^1R$ by

$$ {}^1R' = \left(\frac{\sqrt{R_{min}} - \sqrt{\rho_1}}{\sqrt{\rho_1 R_{min}} - 1} \right)^2 = \frac{R_{min} - 2\sqrt{\rho_1 R_{min}} + \rho_1}{\rho_1 R_{min} - 2\sqrt{\rho_1 R_{min}} + 1}, \tag{A.34}$$

of the specimen:

$$n = \frac{\frac{1}{2}(N_1^2 - 1)}{N_1 \dfrac{(1 + R_{min}) \cdot (1 + \rho_1) \pm 4\sqrt{\rho_1 \cdot R_{min}}}{(1 - R_{min}) \cdot (1 - \rho_1)} - \dfrac{1 + R}{1 - R}}, \tag{10.17}$$

$$k = \sqrt{\frac{R \cdot (n + 1)^2 - (n - 1)^2}{1 - R}} \tag{9.15}$$

R reflectance of the specimen in air (fraction)
R_{min} minimum reflectance (fraction)
ρ_1 reflectance of the layer-air border face (fraction)
N_1 refractive index of the layer material

where $\rho_1 = {}^1R$; $n = \dfrac{\frac{1}{2}(N_1^2 - 1)}{N_1 \dfrac{1 + \rho_1}{1 - \rho_1} - \dfrac{1 + R}{1 - R}}$. \tag{A.37}

Equations (10.17) and (9.15) are shown graphically in Figure A.7.

    Equations (10.17) and (9.15) are ambiguous because of the positive or negative sign of the term $4\sqrt{\rho_1 R_{min}}$ resulting from the positive or negative sign of $\sqrt{R_{min}}$. In general we do not know the sign and we must chose between the two values of n and k by other knowledge; the values are single ones only when R_{min} is zero.

---

for which $\rho > {}^1R$ and the corresponding curves are below the zero curve, or

$${}^1R'' = \left(\frac{\sqrt{R_{min}} + \sqrt{\rho_1}}{\sqrt{\rho_1 R_{min}} + 1}\right)^2 = \frac{R_{min} + 2\sqrt{\rho_1 R_{min}} + \rho_1}{\rho_1 R_{min} + 2\sqrt{\rho_1 R_{min}} + 1}, \tag{A.35}$$

for which $\rho < {}^1R$ and the corresponding curves are above the zero curve, or

$${}^1R''' = \rho_1, \tag{A.36}$$

for which $\sqrt{R_{min}} = 0$, and the zero curve is involved.

R_{min} minimum reflectance (fraction); ρ_1 reflectance of the layer-air border face (fraction) where

$$\rho_1 = \left(\frac{N_1 - 1}{N_1 + 1}\right)^2 \tag{10.18}$$

N_1 refractive index of the layer substance; the value must be given

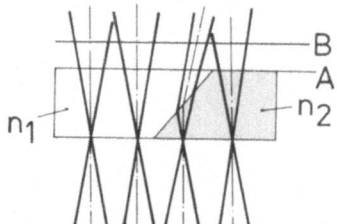

Fig. A.8. Formation of the light (Becke) line at the junction of two contiguous transparent media on change of focus. The figure shows a series of marginal rays intersecting in different points on the medium (n_1) and on medium (n_2) where $n_2 > n_1$. Refraction at glass-air surfaces and cement-medium-surfaces is neglected and we take both glass and cement to have the same refractive index.

We are studying only the refraction effect at the junction of the two media. We take it that the figure shows an even distribution of light over the in-focus level (A) of the stage object in the microscopic image except at the border face. We can see that, if the in-focus level is raised from (A) to (B) the eye will see an extra concentration of light moving into the medium (n_2)

9. Determination of Relative Difference in Refractive Index of Two Contiguous Media

In transmitted light a light line (Becke line) appears at the boundary of media of different refractive index (Fig. A.8). This is due to refraction, except when the size of the two grains approaches the limit of resolution, in which case diffraction is involved; this makes no change because the behaviour of the line is the same in both cases. When the distance between objective and stage object is increased by slowly moving the objective out of its focusing position, the light line appears to move into the medium of higher refractive index. The light line can be seen more clearly with a narrow effective numerical aperture of the condenser.

10. Theory of Secondary Effect of Glare

The secondary effect of glare, causing a systematic error in measurement of specular reflectance is described in Chapter 12, Section 3.1 and here we derive its mathematical expression in two steps. Of course, this is only an approach to the actual conditions.

Step (1): We take the objective lenses and all glass faces situated between the stage object and the reflector to be condensed to a single perfectly transparent layer, normal to the microscope axis having specular reflectance ρ (for vertical incidence) on both the image- and the object-side. Further we take the stage-object surface reflectance (R) to be parallel with this layer and to be parallel with this layer and to lie far enough from the objective so that no interference occurs. The inter-reflections between optics and stage object are illustrated in Figure A.9 and expressed by the following geometrical series:

$$\lim_{n \to \infty} [\rho + (1-\rho)^2 \cdot R(1 + \rho R + \rho^2 R^2 + \cdots + \rho^n R^n)] = \rho + (1-\rho)^2 \cdot \frac{R}{1 - \rho R} \quad (A.38)$$

ρ reflectance of the totality of optics situated between the stage object and the reflector under the operation conditions (fraction)

R reflectance of the stage object (fraction)

Note: In practice ρ is not constant for a given set of glass faces situated between the stage object and the reflector but depends on the operating conditions.

The corresponding measuring value, being proportional to the relative intensity is expressed as

$$G^* \propto \rho + (1-\rho)^2 \frac{R}{1-\rho R} \tag{A.39}$$

G^* measuring value for the total light reflected at the stage object and the objective including the effect of inter-reflections (The asterisk indicates that the measuring value contains an error due to the primary and the secondary effect of glare)

We proceed according to Chapter 12, Section 3.1 and substract from G^* the measuring value for the primary effect of glare, i.e. for the reflectance of glass faces located between the stage object and the reflector ($\Delta G_i \propto \rho$) and

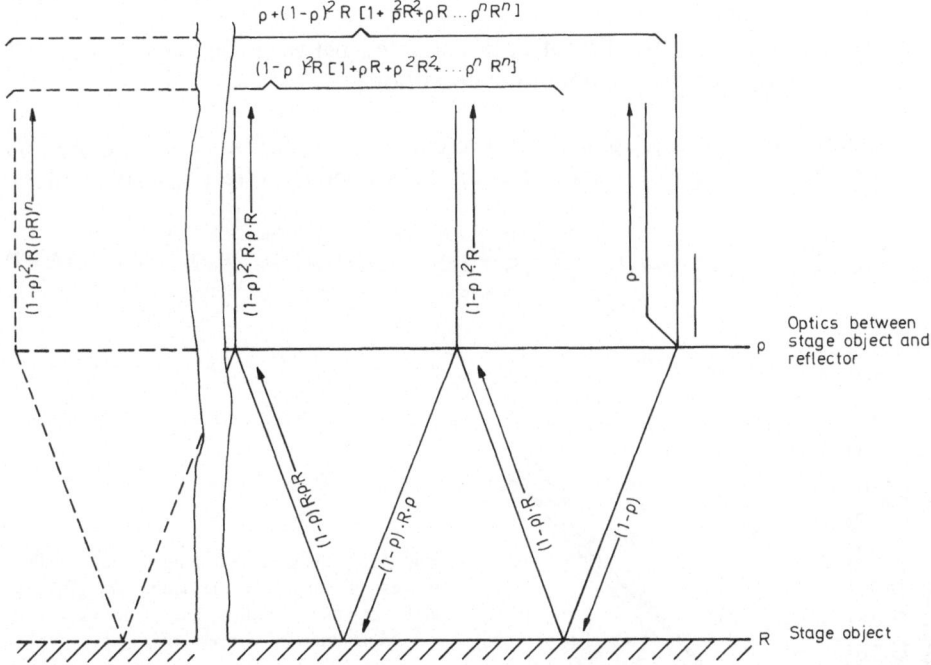

Fig. A.9. Inter-reflections between the stage object and optics situated between the stage object and the reflector for operation with reflected light (schematic). The rays are displaced to show their reflection.

The light is taken as coming from above (the side of the reflector) and as having intensity 1. The surfaces of optics situated between the stage object and the reflector are condensed to a single layer having reflectance ρ on either side; R is the reflectance of the stage object. Term at top of drawing defines the relative intensity in the totality of light beams leaving the optics towards the side of the image obtain

$$G^* - \Delta G_i = G^{**} \propto (1-\rho)^2 \frac{R}{1-\rho R} \tag{A.40}$$

ΔG_i incremental measuring value for the reflectance ρ
G^{**} measuring value corrected in respect to the primary effect of glare

Step (2): We express the measuring values with the primary effect of glare subtracted, for both the specimen and the reference material, by means of Eq. (A.40) and hence the measuring result by

$$\frac{G^{**}_{sp}}{G^{**}_{ref}} \cdot R_{ref} = R^{**}_{sp} = \frac{R_{sp}(1-\rho R_{ref})}{R_{ref}(1-\rho R_{sp})} \cdot R_{ref} = R_{sp}\frac{(1-\rho R_{ref})}{(1-\rho R_{sp})} \tag{A.41}$$

R^{**}_{sp} measuring result (fraction), containing within it the error due to the secondary effect of glare
R true reflectance (fraction)
sp specimen
ref reference material
ρ reflectance of the totality of optics situated between the stage object and the reflector under the operating conditions

Equation (A.41) is graphically represented in Figure A.10 taking R_{ref} as zero or one. The absolute error due to the secondary effect is expressed by

$$R^{**}_{sp} - R_{sp} = \Delta R_{sp} = R_{sp} \cdot \rho \cdot \frac{R_{sp} - R_{ref}}{1-\rho R_{sp}} \tag{A.42}$$

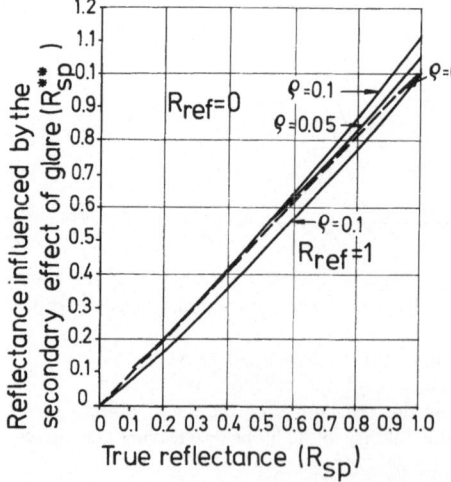

Fig. A.10. Diagram showing specular reflectances influenced by the secondary effect of glare for various reflectances of optics.
Mathematical expression of the graphs:

$$\frac{G^{**}_{sp}}{G^{**}_{ref}} \cdot R_{ref} = R^{**}_{sp} = R_{sp} \frac{1-\rho R_{ref}}{1-\rho R_{sp}}. \tag{A.41}$$

The graphs refer to $R_{ref}=0$ together with $\rho=0.05, 0.1$, and $R_{ref}=1$ together with $\rho=0.1$.

Note: The absolute error $(R^{**}_{sp} - R_{sp})$ is shown by the difference in height between the curves and the diagonal and the relative error by

$$\frac{\Delta R_{sp}}{R_{sp}} = \rho \frac{R_{sp} - R_{ref}}{1 - \rho R_{sp}}. \qquad (A.42\,a)$$

If $\rho \cdot R_{sp}$ is extremely small in relation to R_{sp}, this approximates to

$$\frac{\Delta R_{sp}}{R_{sp}} \approx \rho(R_{sp} - R_{ref}) \qquad (A.43)$$

so that the error is proportional to the reflectance of the optics (ρ) and to the difference in reflectance between the specimen and the reference material $(R_{sp} - R_{ref})$; if $\rho = 0$ or $R_{sp} = R_{ref}$ the error is zero.

11. Mathematical Derivation of the Distributional Error

A non-uniform field is occupied by materials of transmittance T_1 and T_2 respectively. For the determination of the mean absorbance we measure the transmittances T_1 and T_2 independently. Then we express the values in terms of absorbance A_1 and A_2. Through our knowledge of p (the fraction of the field occupied by the first material) we can sum the absorbances

$$p \cdot A_1 + (1-p)\, A_2 = \bar{A} \text{ (true mean absorbance)} \qquad (A.44)$$

Now we put $A = -\lg T$ and obtain

$$\bar{A} = p\, (\lg T_2 - \lg T_1) - \lg T_2 \qquad (A.45)$$

After the independent measurement of T_1 and T_2 it is erroneous to calculate the mean transmittance as

$$\bar{T} = p \cdot T_1 + (1-p) \cdot T_2 = p(T_1 - T_2) + T_2 \qquad (A.46)$$

and to convert the mean transmittance into a 'mean' absorbance as

$$-\lg \bar{T} \equiv \bar{A}^* \qquad (A.47)$$

A absorbance
T transmittance (fraction)
      The subscripts 1 and 2 indicate the material, the bar indicates the true mean value, the bar plus asterisk indicates the mean value containing within it the distributional error.
p fraction of field occupied by the material (T_1), hence $(1-p)$; fraction of field occupied by material (T_2)

Note: The latter procedure means that first the mean transmittance ($\bar{T}$) is determined with a single measurement in the whole field occupied by the materials T_1 and T_2 so that both materials are measured together and then the mean transmittance is transformed into absorbance.

The difference between Eqs. (A.47) and (A.45) is called 'distributional error.' It is expressed as

$$\bar{A}^* - \bar{A} = \Delta\bar{A} = p(\lg T_1 - \lg T_2) + \lg T_2 - \lg[p(T_1 - T_2) + T_2].$$
(A.48)

If we take the material (T_1) as lying in a clear perfectly transmitting field, we can take T_2 as being 1, thus $A_2 = -\lg T_2 = 0$ and Eq. (A.48) becomes

$$\Delta\bar{A} = p \cdot \lg T_1 - \lg[p(T_1 - 1) + 1]$$
(A.49)

hence the 'relative distributional error' becomes

$$\frac{\Delta\bar{A}}{\bar{A}} = f(p, T_1) = \frac{p \cdot \lg T_1 - \lg[p(T_1 - 1) + 1]}{p \cdot \lg T_1}.$$
(A.50)

This function is shown in Figure A.11 where several values of T_1 are taken as parameters.

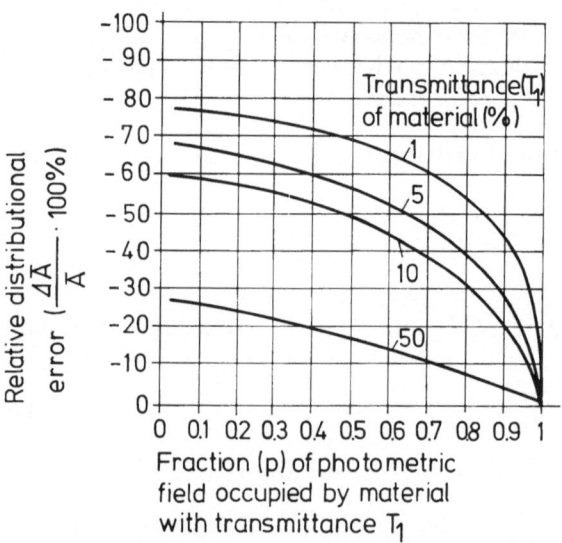

Fig. A.11. Diagram showing relative distributional error for material in a perfectly transmitting background plotted as a function of the fraction of the field occupied by the material.

Defining expression:

$$\frac{\Delta\bar{A}}{\bar{A}} = f(p, T_1) = \frac{p \cdot \lg T_1 - \lg[p(T_1 - 1) + 1]}{p \cdot \lg T_1}$$
(A.50)

$\dfrac{\Delta\bar{A}}{\bar{A}} = \dfrac{\bar{A}^* - \bar{A}}{\bar{A}}$ Relative distributional error; $\bar{A}^*$ absorbance corresponding to the average trans-

mittance in the non-uniform field [see Eq. (A.47)]; $\bar{A}$ true mean absorbance [see Eq. (A.41)]; p fraction of photometric field occupied by the material; T_1 transmittance of material (fraction)

12. Beam Splitting in an Inclined Glass Plate and the Effect of Parasitic Images

An inclined glass plate situated in the imaging optical train splits the light beam in the way as shown in Figure A.12a. There are two cases:

1. The light beam from the stage object has parallel marginal rays so that the glass plate is hit only by parallel rays. These are made to intersect in the primary-image plane with the help of a tube lens (Fig. A.12a).

2. The light beam has marginal rays converging towards the primary image and intersecting in the primary-image plane (Fig. A.12b).

In case (1) both the rays of the component forming the regular image and those of the split-off component intersect in the same point on the primary-image plane due to the effect of the tube lens; the regular image is formed at the same spot at which it would be formed if the glass plate is removed and no image is produced. We need not discuss this case further.

In case (2) the regular image is displaced from the position that it would have without the glass plate and the parasitic beam forms a parasitic image. The lateral distance between the regular and the parasitic image is expressed by

$$d = \frac{t \cdot \sin(2v)}{\sqrt{N^2 - \sin^2 v}}. \tag{A.51}$$

(For meaning of symbols see Fig. A.12.)

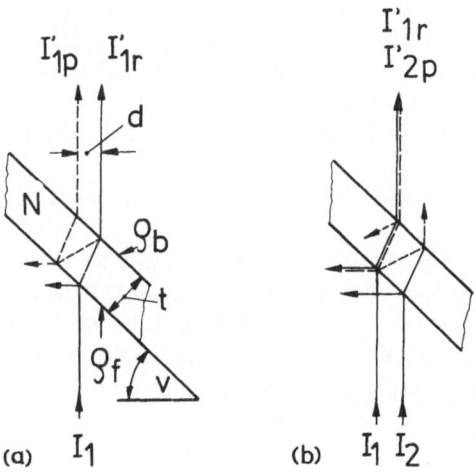

Fig. A.12a and b. Parameters involved in beam splitting and inter-reflection at the front and back of an inclined glass plate. *Full lines:* regular light beam; *dashed lines:* parasitic light beam.

Parameters: ρ reflectance at the surface of the glass plate (fraction) (The reflectance at a surface is taken as being the same for light incident from the object and from the image side). f front (lower or object side), b back (upper or image side); v angle of inclination, subscripts 1 and 2 refer to two different areas in the stage object. d distance between the regular and parasitic beam; t thickness of glass plate; N refractive index of glass plate; I Intensity of light in the stage object (any radiometric and photometric quantity); I' intensity of light in the image. The numbers 1 and 2 indicate the spots in the stage object. p parasitic, r regular.

(a) Splitting and reflection of a beam coming from a single area (I_1) in the stage object. (b) Superposition of two beams in the primary image, regular beam from area (I_1) and parasitic beam from area (I_2)

Practical example: The distance of the parasitic image produced by a glass-plate reflector in the reflected-light microscope of standard design having $t=0.7$ mm, $N=1,5$ and $v=45°$ is $d=0.53$ mm.

Note: The internal reflections are, theoretically, infinitely repeated, but we neglect all except the first-order effects because of the low intensities involved.

Now we study the effect of the superposition of a parasitic image on the regular image. We take the regular image as originating from an area (I_1) and the parasitic one as originating from an adjacent area (I_2) in the stage object (Fig. A.12b). We take these areas as having different intensities (I_1) and (I_2) respectively. If we neglect absorption of light and take the reflectances of the front face and of the back face of the glass plate as being the same for either direction of light, we can express the intensity in the regular image of spot (I_1) by

$$I'_{1r}=I_1(1-\rho_b)(1-\rho_f) \tag{A.52}$$

and the intensity in the parasitic image of spot (I_2) by

$$I'_{2p}=I_2(1-\rho_b)(1-\rho_f)\cdot\rho_b\cdot\rho_f \tag{A.53}$$

(For the meaning of symbols see Fig. A.12.)

In the area where the parasitic image of spot (I_2) is superposed on the regular image of spot (I_1) the integral intensity is

$$I'_{1r}+I'_{2p}=(1-\rho_b)(1-\rho_f)(I_1+I_2\cdot\rho_b\cdot\rho_f). \tag{A.54}$$

Actually, if the diameter of the illuminated area in the primary image in the direction d (Fig. A.12a) is at least twice the distance d [Eq. (A.51)] every point on the image field is hit by a regular and a parasitic beam. Although a parasitic image may be present it is seen only when two conditions are obeyed.

1. The area (I_2) from which it comes must be brighter than that from which the regular image comes (I_1).

2. The diameter of the image of area (I_1) in direction d must be greater than d.

Now we describe the measuring error caused by the parasitic image. First we take area (I_1) as being occupied by the specimen and area (I_2) by a feature that does not belong to our specimen. We replace the subscript 1 by sp and maintain the subscript 2 for the feature and express the intensities supplied by specimen and the feature in terms of measuring values, thus $G_{sp}\propto I_{sp}$, and $G_2\propto I_2$

G hypothetical measuring value
I intensity (any radiometric or photometric quantity)
sp specimen
2 feature lying by the side of the specimen in the direction d

Note: The measuring values are hypothetical because they cannot be displayed separately when the superposition occurs. They can only be displayed by means of a separate test, in which the glass plate is removed from the optical train and the specimen and the feature is put one after the other into the photometric beam.

We replace the symbols I_1 and I_2 in Eqs. (A.52) and (A.53) by G_{sp} or G_2 respectively and express the sum of intensities $(I'_{1r}+I'_{2p})$ of Eq. (A.54) by the corresponding measuring value (G^*_{sp}) so that

$$G^*_{sp}=(1-\rho_b)(1-\rho_f)(G_{sp}+G_2\cdot\rho_b\cdot\rho_f). \tag{A.55}$$

Then we replace the specimen together with the feature by a homogeneous reference material the area of which is so large that both areas (I_1) and (I_2) are lying in the same material. In this case the intensities I_1 and I_2 are equal so that the hypothetical measuring values for the areas (I_1) and (I_2) can be expressed by

$$G_{ref}=G_2\propto I_{ref}=I_2$$

and the measuring value for the sum of intensities of the regular image of area (I_1) and the parasitic image of area (I_2) by

$$G^*_{ref}=(1-\rho_b)(1-\rho_f)(G_{ref}+G_{ref}\cdot\rho_b\cdot\rho_f). \tag{A.56}$$

The quotient of Eqs. (A.55) and (A.56) is our measuring result, thus

$$\frac{G^*_{sp}}{G^*_{ref}}=\frac{G_{sp}+G_2\cdot\rho_b\cdot\rho_f}{G_{ref}(1+\rho_b\rho_f)}, \tag{A.57}$$

while the correct result, i.e. the result obtained without the influence of parasitic images, is G_{sp}/G_{ref}.

We express the absolute error by

$$\frac{G^*_{sp}}{G^*_{ref}}-\frac{G_{sp}}{G_{ref}}=\Delta\left(\frac{G_{sp}}{G_{ref}}\right)=\frac{\rho_b\cdot\rho_f(G_2-G_{sp})}{G_{ref}(1+\rho_b\cdot\rho_f)} \tag{A.58}$$

and the relative error by

$$\frac{\Delta\left(\dfrac{G_{sp}}{G_{ref}}\right)}{\dfrac{G_{sp}}{G_{ref}}}=\frac{\rho_b\cdot\rho_f(G_2-G_{sp})}{G_{sp}(1+\rho_b\cdot\rho_f)} \tag{A.59}$$

Meaning of symbols in Eqs. (A.55) to (A.59);
G measuring value
I intensity of light (any photometric or radiometric quantity)
sp   specimen

2   feature lying by the side of the specimen and giving rise to the parasitic
        image
ref   reference material
ρ reflectance of the glass plate (fraction)
f   front side
b   back side

The asterisk indicates that the measuring value is influenced by the parasitic
image.

13. Derivation of the $\cos^4$-Law

In Chapter 3, Section 4.1 we have stated that the optical flux in a centred
narrow light tube that supplies small angles of incidence of light on either
emitting area is expressed by

$$OF = \frac{F_1 \cdot F_2}{l^2} \tag{A.60}$$

F_1, F_2 areas limiting the light tube on either side
l distance between the area along the optical axis

Now we take the light tube as being inclined in respect to the axis of the
optical system but the limiting surfaces as being normal to the axis of the
system, hence as being oblique in respect to the axis of the light tube (Fig. A.13a).

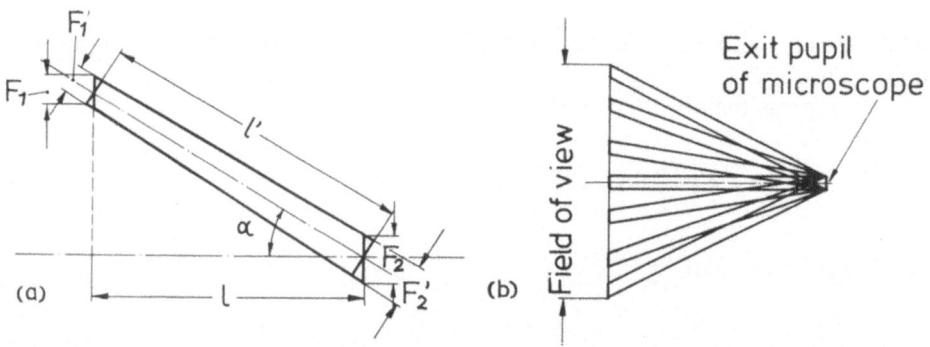

Fig. A.13a and b. Optical flux for oblique directions.

(a) Parameters determining the optical flux in an oblique light tube. F_1, F_2 areas of limiting
faces; l distance of limiting faces in the direction normal to the faces; F_1', F_2' effective areas
of limiting faces in the direction of the tube axis; l' distance of the centres of the limiting faces
in the direction of the tube axis; α angle subtended between the axis of the light tube and
the normal of the limiting faces.

(b) Longitudinal section through the light train in the microscope between the apparent field
in the ocular and the exit pupil of the microscope showing that this train can be subdivided
into a series of partial light tubes with varying angular distances from the microscope axis, the
angles being field-of-view angles ($w/2 \equiv \alpha$). For each partial tube Eq. (A.62) is applied

The inclination is taken as being so great that the angle of incidence of light on the limiting surfaces must not be neglected. In this case we have to convert the actual areas of the limiting surfaces and their actual distance along the axis of the optical system into effective areas and an effective distance. Using the effective parameters, the optical flux in the inclined tube is expressed formally in the usual way by

$$OF = \frac{F_1' \cdot F_2'}{l'^2} \tag{A.61}$$

F_1', F_2' effective limiting areas on either end of an oblique light tube
l' distance along the normal of the areas

Taking the angle subtended between the axes of the light tube and the optical system as being α, the effective parameters are related to the actual ones (F_1, F_2, l) by the factor $\cos \alpha$ in the form

$$F_1' = F_1 \cdot \cos \alpha, \ F_2' = F_2 \cdot \cos \alpha, \ l' = l/\cos \alpha$$

(see Lambert's cosine law, Fig. 3.1 b) so that Eq. (A.61) is expressed by

$$OF = \frac{F_1 \cdot F_2 \cdot \cos^4 \alpha}{l^2}. \tag{A.62}$$

Note: Although the $\cos^4$-law is mainly important in the field of macrophotography, where it dictates the distribution of brightness across a photographic image, this law can, with a certain restriction, also dictate the diminution of light intensity from the centre to the edge of a field in the microscope, especially if the field in a wide-angle ocular is concerned. This is seen if the field is subdivided in several spots lying at different distances from the centre of the field or having different angular distances from the microscope axis (Fig. A.13 b). In this case the angle is a half of a field angle ($\alpha = w/2$).

14. Determination of Angles Between the Microscope Axis and Rays Deflected at, or Traversing, the Stage Object

In the measurements here described, a micrometer ocular is used throughout and the image of the aperture area of the objective is viewed through the Bertrand lens (Chap. 2, Sect. 3.2).

The relation between angles and central distances is illustrated in Figure A.14. The most important angles are the aperture angle (effective and potential) and the obliquity of illumination. (See also Fig. 2.1 e and f.)

By comparing a known aperture angle or the corresponding known numerical aperture with the central distance on the micrometer scale we can calibrate the scale in terms of angle (or numerical aperture). Taking, for example (Fig. A.14a), the distance on the micrometer scale between the centre of the scale and the margin of the image of the aperture area to be 10 scale intervals

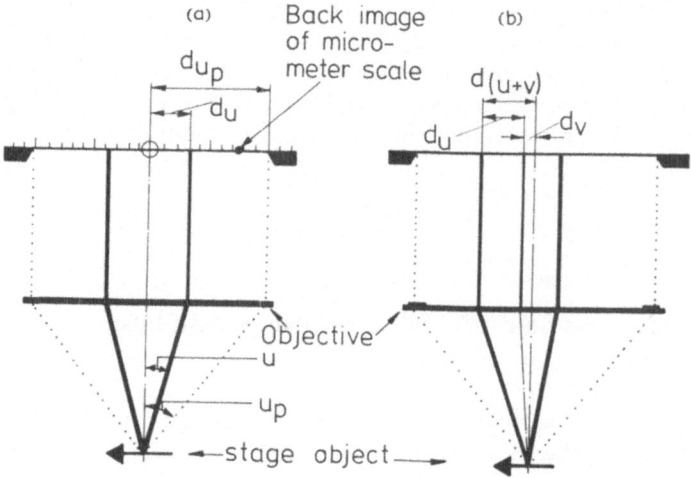

Fig. A.14a and b. Relation between central distances in the aperture area of the objective and angles subtended at the stage object.

(a) Normal bright-field illumination. (b) Oblique bright-field illumination. u effective aperture angle; u_p potential aperture angle; v obliquity, i.e. angle between the axes of the objective and of the effective light cone. [This angle is not indicated by the symbol in Figure (b)]; $u+v$ maximum angle of incidence; d central distance; the subscript indicates the corresponding angle.

Note: In oblique bright-field illumination the aperture angle (u) is obtained from the relation $d_u = d_{(u+v)} - d_v$ where $d_{(u+v)}$ and d_v are measured

and the potential numerical aperture of the objective to be 0.6 the calibration constant (Mallard's constant) is 0.06 units of sin α per scale interval.

An almost linear relationship exists between the sine of angles and central distances. A more accurate relation is obtained by comparing a series of known angles with the corresponding central distance with the help of an aperture disc or an apertometer, as are described below.

Notes:

1. The calibration is only true for a given set of optics (Bertrand lens, micrometer scale, and ocular) and for a constant focusing position of the Bertrand lens.

2. A sharp image of the condenser-aperture diaphragm is only achieved if the stage object is clear in transmitted light, or has a flat and homogeneous specularly reflecting surface in reflected light.

3. In an immersion liquid with refractive index N, or as an internal angle in a transparent stage object with refractive index n the angle (α) is given by sin $\alpha = \sin \alpha'/N$ or sin $\alpha = \sin \alpha'/n$, where α' is the angle in air. From this we can draw the important conclusion that, for a given numerical aperture, the maximum angle of incidence with an immersion objective is always smaller than that with a dry objective (Chap. 6, Sect. 4.1).

4. In polarising microscopy, by this means the central distances of certain features in 'conoscopic interference images' are measured and the angles between the corresponding cristal-optical directions (e.g. the optic-axial angle) determined.

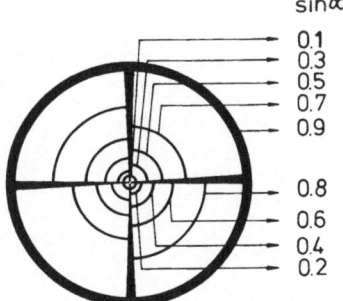

sinα

0.1
0.3
0.5
0.7
0.9

0.8

0.6

0.4
0.2

Fig. A.15. Volkmann's aperture disc (half real diameter)

Volkmann's Aperture Disc

This can only be used if it is possible to place the disc at a distance of 25 mm below the front focal plane of the objective. The disc is made from any stiff material (Fig. A.15) and it must be placed in this position. It is illuminated by an auxiliary light source so that its image is observed through the Bertrand lens in the back focal plane of the objective; the disc is moved laterally until its image is centred.

The disc shows segments of circles that are indicated in terms of sin α. By relating the set of circles to the central distances of points where the segments intersect the ocular-micrometer scale we can directly calibrate the corresponding set of central distances on the scale in terms of sin α. This method of calibration is superior to that shown in Figure A.14a because deviations from the linear relation between central distances and sin α are taken into account.

The disc can be drawn for any set of sin α, the radii of the circle segments being

$$r = \frac{2.5 \text{ mm} \cdot \sin \alpha}{\sqrt{1 - \sin^2 \alpha}} \tag{A.63}$$

hence, for example,

sin α	0.1	0.2	0.3	0.4	0.5	0.6	0.7	0.8	0.9
r(mm)	2.5	5.1	7.9	10.9	14.4	18.8	24.5	33.3	51.6

Notes:

1. It is not possible to use the disc with oil immersion.

2. For an objective corrected for an infinite image distance the focal plane on the stage-object side is the same as the stage-object plane. For an objective with correction for a finite distance, the focal plane lies somewhat closer to the objective than the stage-object plane.

3. A modification of the shape of the disc looking like a cat's tongue is called Aperture disc after C. Metz.

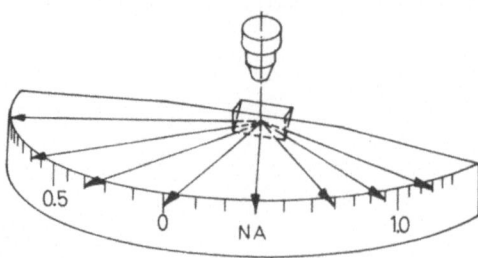

Half-Disc Apertometer

The apertometer is made only by a few microscope manufacturers. The figure (Fig. A.16) shows the principle of its technical design.

The apertometer is placed on the microscope stage. The objective is focused on a mark on the centre of the apertometer and the peripheral scale is illuminated so that an image of the scale is seen in the aperture area of the objective. For calibration of the micrometer scale we have to read the length at which it is intersected by the (sin α)-lines of the apertometer-scale image, and to plot sin α versus central distance on the scale.

Note: The use of the apertometer allows the highest accuracy in calibration to be achieved, including that for oil-immersion objectives.

Mathematical Relation Between Central Distance and Characteristic Parameters of the Objective

For small aperture angles, the central distance in the aperture area is expressed by

$$u \approx \sin u \approx \tan u = d/f \tag{A.64}$$

u aperture angle (in rad)
d central distance corresponding to u
f focal length of the objective (in the same unit as d)

In Equation (6.7) we have $f = f_T/MN_{Obj}$. We can now express the aperture angle in terms of magnification number

$$u \approx d \cdot MN_{Obj}/f_T \tag{A.65}$$

MN_{Obj} magnification number of the objective
f_T tube length of the microscope

After rearranging Eq. (A.65) it is evident that, for a given angle (u) with a given tube length (f_T) the central distance is inversely proportional to the magnification number,

$$d \approx u \cdot f_T / M N_{Obj} \qquad\qquad (A.66)$$

($u =$ constant)
($f_T =$ constant)

Note: If we take u as being the angle with which a beam incident on a specularly reflecting and inclined surface is deflected from the direction of the microscope axis we can take $u/2$ as being the angle of tilt of the surface, hence apply Eq. (A.66) also to the angle of tilt. In this case, d is the central distance of the centre of the image of the condenser-aperture diaphragm (Chap. 8, Sect. 4), provided that the centre of the image perfectly coincides with the centre of the aperture area, when the surface is perfectly levelled.

Note added in proof

In the time when the proofs were corrected the author has learnt of the following paper on systematic errors. He considers it important for the subject of the book (pp. 161–188): Reule, A.G.: Errors in spectrophotometry and calibration procedures to avoid them. Journal of Research N.B.S. (USA) — A. Physics and Chemistry **80A**, pp. 609–624 (1976).

References

Adams, J.W.: The visible absorption spectra of rare-earth minerals. Am. Mineralogist **50**, 356–366 (1965).

Allen, R.D., Brault, J.W., Zeh, R.M.: Image contrast and phase-modulated light methods in polarization and interference microscopy. In: Advances in Optical and Electron Microscopy (eds. R. Barer, V.E. Coslett), Vol. 1, pp. 77–114. London: Academic Press 1966.

Alpern, B., Correia, M., van Gijzel, P., Jacob, H., Teichmüller, M., Wolf, M.: Fluorescence microscopy and fluorescence photometry. In: International Vocabulary for Coal Petrology Suppl. 1975 (ed. International Commission for Coal Petrology, Analysis Subcommission). Paris: C.N.R.S. 1975.

Anders, H.: Dünne Schichten für die Optik. Stuttgart: Wissenschaftliche Verlagsges. 1965.

Anex, B.G.: Optical properties of highly absorbing crystals. Mol. Crystals **1**, 1–36 (1966).

Bahr, G.F.: Determination of dry mass of small biological objects by quantitative electron microscopy. In: Micromethods in Molecular Biology (ed. V. Neuhoff), pp. 257–284. Berlin-Heidelberg-New York: Springer 1973.

Barer, R.: Phase contrast and interference microscopy in cytology. In: Physical Techniques in Biological Research (eds. G. Oster, A.W. Pollister), Vol. III, pp. 30–90, New York: Academic Press 1956.

Bartels, P.H.: Sensitivity and evaluation of microspectrophotometric and microinterferometric measurements. In: Introduction to Quantitative Cytochemistry (ed. G.L. Wied), pp. 93–105. New York-London: Academic Press 1966.

Beilby, G.T.: Aggregation and Flow of Solids. London: Macmillan and Co. 1921.

Beneke, G.: Application of interference microscopy. In: Introduction to Quantitative Cytochemistry (ed. G.L. Wied), pp. 63–92. New York-London: Academic Press 1966.

Berek, M.: Ein Prisma für 90°-Ablenkung, bei dem die Störungen im Polarisationszustand eines wenig geöffneten räumlichen Strahlenbündels korrigiert sind. Z. Instrumentenkunde **56**, pp. 1–6 (1936).

Bessmertnaja, M.S., Tshvileva, T.N., Agroskin, L.S., Botchev, L.J., Lebedeva, S.M., Poginova, L.A.: The determination of ore minerals in polished sections by reflectance spectra and hardness (in Russian). Moscow: Nedra 1973.

Beyer, H.: Theorie und Praxis der Interferenzmikroskopie. Leipzig: Akademische Verlagsges. 1974.

Billmeyer, F.W., Jr., Saltzmann, M.: Principles of color technology. New York-London-Sydney: Interscience Publishers 1966.

Blout, E.R., Bird, G.R., Grey, D.S.: Infrared microspectroscopy. J. Opt. Soc. Am. **40**, 304 (1950).

Boguth, W.: Scanning-Interferenzmikroskopie mit dem Mikroskop-Photometer. Microscopica Acta **76**, pp. 28–37 (1974).

Bowie, S.H.U., Simpson, P.R., Atkin, D.: Reflectance measurements in monochromatic light on the Bowie-Taylor suite of 103 ore minerals. Fortschr. Miner. **52**. Spec. Issue: IMA-Papers 9th Meeting, Berlin-Regensburg 1974, pp. 567–582 (1975).

Brockhaus ABC der Optik (ed. K. Mütze, F. Foitzik, W. Krug, G. Schreiber). Leipzig: VEB Brockhaus Verlag 1961.

Burnett, R.W.: Accurate measurement of molar absorptivities. J. Res. Nat. Bur. Stand (U.S.) **76 A**, p. 483 (1972).

Burns, R.G., Vaughan, D.J.: Interpretation of the reflectivity behaviour of ore minerals. Am. Mineralogist **55**, pp. 1576–1586 (1970).

Carlson, L., Caspersson, T.C., Lomakka, G., Silverbåge, S.: A rapid and integrating microinterferometer for large-scale population work. In: Introduction to Quantitative Cytochemistry — II (eds. G.L. Wied, G.F. Bahr), pp. 117–123. New York-London: Academic Press 1970.

Caspersson, T.C., Lomakka, G.: Recent progress in quantitative cytochemistry: Instrumentation and results. In: Introduction to Quantitative Cytochemistry – II (eds. G.L. Wied, G.F. Bahr), pp. 27–56. New York-London: Academic Press 1970.

Cervelle, B., Lévy, C.: Dosage rapide du magnésium dans les ilménites par microréflectométrie. Mineralium Deposita (Berlin) **6**, pp. 34–40 (1971).

Cheznokov, B.V.: Spectral absorption curves for certain minerals colored by titanium (in Russian). Oklady Acad. Sci. USSR. Earth Sci. Sect. **129**, pp. 1162–1164 (1960).

Claussen, H.C.: Mikroskope. In: Encyclopedia of Physics (ed. J. Flügge), Vol. XXIX, pp. 343–425. Berlin-Heidelberg-New York: Springer 1967.

David, G.D., Galbraith, W.: The Denver universal microspectroradiometer (DUM) I. General design and construction. J. Microsc. **103**, pp. 135–178 (1975).

Davies, H.G.: Microscope interferometry. In: General Cytochemical Methods (ed. I.F. Danielli), Vol. 1, pp. 57–101. New York: Academic Press 1958.

Davis, A., Vastola, F.J.: Developments in automated reflectance microscopy of coal. J. Microsc. In press (1977).

Deitsch, A.D.: Cytophotometry of proteins. In: Introduction to Quantitative Cytochemistry (ed. G.F. Wied), pp. 451–468. New York-London: Academic Press 1966.

Dörmer, P.: Quantitative autoradiography at the cellular level. In: Micromethods in Molecular Biology (ed. V. Neuhoff), pp. 347–393. Berlin-Heidelberg-New York: Springer 1973.

Edström, J.E.: Micro-electrophoresis for RNA and DNA base analysis. In: Micromethods in Molecular Biology (ed. V. Neuhoff), pp. 215–256. Berlin-Heidelberg-New York: Springer 1973.

Evans, D.M. (ed.): Cytology Automation (Proc. of the 2nd Tenovus Symp. Cardiff, 1968). Edinburgh-London: Livingstone 1970.

Ewing, G.W.: Instrumental Methods of Chemical Analysis. New York-Toronto-London: McGraw-Hill 1954.

Exner, G., Schreiber, W.: Mikrophotometrische Untersuchungen an Modellpräparaten. In: Optik und Spektroskopie aller Wellenlängen (ed. P. Görlich), p. 236. Berlin: Akademie-Verlag 1962.

Faye, G.H., Nickel, E.H.: On the origin of colour and pleochroism of kyanite. Can. Mineralogist **10**, pp. 30–46 (1969).

Fisher, R.A.: Statistical Methods for Research Workers. Edinburgh-London: Oliver and Boyd 1958.

Flügge, J.: Leitfaden der geometrischen Optik und des Optikrechnens. Göttingen: Vandenhoek und Rupprecht 1956.

Forsyth, W.E., Worthing, A.G.: The properties of tungsten and the characteristics of tungsten lamps. Astrophys. J., vol. LXI, pp. 146–185 (1925).

Françon, M.: Progress in Microscopy. Oxford-London-New York-Paris: Pergamon Press 1961.

Freed, J.J.: Microspectrophotometry in the ultraviolet spectrum. In: Physical Techniques in Biological Research (ed. A.W. Pollister), vol. III C, 2nd ed. New York: Academic Press 1969.

Frei-Sulzer, M.: Coloured fibres in criminal investigation with special reference to natural fibres. In: Methods of Forensic Science, vol. IV, pp 141–176. London: John Wiley and Sons 1965.

Galbraith, W., Geyer, S.B., David, G.B.: The Denver universal microspectroradiometer (DUM), II. Computer configuration and modular programming for radiometry. J. Microsc. **105**, pp. 237–264 (1975).

Galopin, R., Henry, N.F.M.: Microscopic Study of Opaque Minerals. Cambridge: Heffer 1972.

Gijzel, P. van: Polychromatic UV-fluorescence microphotometry of fresh and fossil plant substances, with special reference to the location and identification of dispersed organic material in rocks. In: Colloque internationale. Pétrographie de la matière organique des sédiments, relations avec la paléotempérature et le potentiel pétrolier, pp. 67–91. Paris: C.N.R.S. 1973.

Gijzel, P. van, Schwirtlich, J.: Determination, correction and evaluation of fluorescence parameters by means of the computer-operated microscope-spectrophotometry. J. Microsc. In press (1977).

Gilbert, L.A.: The reflectivity of coal vitrains in the visible and the ultra-violet. Fuel **40**, pp. 72–73 (1961).

Goldstein, D.J.: Aspects of scanning microdensitometry. I. Stray light (glare). J. Microsc. **92**, pp. 1–16 (1970). II. Spot size, focus and resolution. J. Microsc. **93**, pp. 15–42 (1971). III. The monochromator system. J. Microsc. **105**, pp. 33–56 (1975).

Graf, U., Henning, H.-J., Stange, K.: Formeln und Tabellen der mathematischen Statistik, 2nd ed. Berlin-Heidelberg-New York: Springer 1966.

Green, D.K.: Chromosome analysis. In: Cytology Automation, Proc. 2nd Tenovus Symp. Cardiff, 1968 (ed. D.M.D. Evans). Edinburgh-London: Livingstone 1970.

Gregory, R.L.: Eye and Brain (Reprinted 2nd ed.). London: World University Library, Weidenfeld and Nicolson 1973.

Greif, H.: Lichtelektrische Empfänger. Leipzig: Akademische Verlagsges. 1972.

Grundmann, E., Hobik, H.P., Schühly, A.: Die Mikrospektrophotometrie der Zelle im sichtbaren Spektralbereich. Deut. Med. Wochenschr. **88**, pp. 1–16 (1963).

Hansen, G.: Abbildung von Volumenstrahlern in Spektrographen. Optik **6**, pp. 337–347 (1950).

Hardy, A.C.: Handbook of Colorimetry. Cambridge, Mass: The Technology Press 1936.

Helwig, H.-J.: Neue Standard Glühlampe. Lichttechnik **1**, pp. 110–111 (1949).

Henry, N.F.M., Phillips, R.: Quantitative Colour and its Use in Microscopic Mineralogy. London: McCrone Research Associates, Ltd. 1977.

Herz, R.H.: Photographic aspects of X-ray crystallography. In: X-ray Diffraction by Polycrystalline Materials (eds. H.S. Pliser, H.P. Rooksby, A.J.C. Welson). London: Institute of Physics 1955.

Hoel, P.G.: Introduction to mathematical statistics. 4th ed. New York-London: John Wiley 1971.

Höfert, H.-J.: The "speed" (Lichtstärke) of monochromators. Zeiss Information No. **60**, pp. 59–64 (1966).

Howling, D.H., Fitzgerald, P.J.: The nature, significance and evaluation of Schwarzschild-Villiger effect in photometric procedures. J. Biophys. Biochem. Cytol. **6**, pp. 313–337 (1959).

Hummel, R.E.: Optische Eigenschaften von Metallen und Legierungen. Berlin-Heidelberg-New York: Springer 1971.

Jackson, W.A., Hodgson, W.A., Wallen, V.R., Philpotts, L.E., Hunter, J.: Potato late blight intensity levels as determined by microdensitometer studies of false colour aerial photographs. J. Biol. Photogr. Assoc. **39**, pp. 101–106 (1971).

Jacob, H., Knickrehm, E.: Umrechnung des mikroskop-photometrisch ermittelten Reflexionsvermögens von Feststoffen für beliebige Immersionsflüssigkeiten. Conversion for various immersion fluids of microphotometrically determined reflectivities of solids. Microscopica Acta **77**, pp. 301–308 (1975).

Jenkins, F.A., White, H.E.: Fundamentals of Optics, 3rd ed. New York: McGraw-Hill 1957.

Kaiser, H., Specker, H.: Bewertung und Vergleich von Analysenverfahren. Z. Analyt. Chemie **149**, pp. 46–66 (1956).

Klaunig, W.: Der Lichtleitwert. Feingerätetechnik **2**, p. 179 (1953).

Knosp, H.: Microphotometry – Uses in Metallography. Prac. Metallogr. **7**, pp. 494–509 (1970) (English and German).

Konev, S.V.: Fluorescence and Phosphorescence of Proteins and Nucleic Acids. New York: Plenum Press 1967.

Koritnig, S.: Das Reflexionsvermögen opaker Mischkristallreihen. N. Jb. Mineral. (Monatsh.) **8**, pp. 225–231 (1964).

Koritnig, S.: Zur Bestimmung der Brechzahl nicht- oder schwach absorbierender Medien aus dem Reflexionsvermögen. Archiv für Lagerstättenforschung in den Ostalpen. Special Issue **2**, pp. 153–159 (1974).

Kortüm, G.: Kolorimetrie, Photometrie und Spektrometrie. 4th ed. Berlin-Göttingen-Heidelberg: Springer 1962.

Kortüm, G.: Reflexionspektroskopie. Berlin-Heidelberg-New York: Springer 1969.

Kranert, W., Raether, H.: Über die Strukturänderung von Metallen durch Kaltbearbeitung. Annal. Physik **43**, pp. 520–537 (1943).

Krug, W., Rienitz, J., Schulz, G.: Beiträge zur Interferenzmikroskopie. Berlin: Akademie Verlag 1961.

Ku, H.H.: Notes on the use of propagation of error formulas. J. Res. Nat. Bur. Stand (U.S.) **70 C**, p. 263 (1960).

Ku, H.H. (ed.): Precision measurement and calibration. Statistical conceps and procedures: National Bureau of Standards. Special Publication **300**, vol. 1, 1969.

Landolt-Börnstein: Zahlenwerte und Funktionen aus Physik, Chemie, Astronomie, Geophysik, Technik. Optische Konstanten (Optical Constants), Vol. I, pt. 3, Vol. II, pt. 8, 1962. Berlin-Göttingen-Heidelberg: Springer 1951.

Langer, K., Abu-Eid, R.M.: Measurement of the polarized absorption spectra of synthetic transition metal-bearing silicate micro-crystals in the spectral range 44000–4000 cm^{-1} (227–2500 nm). Physics and Chemistry of Minerals. In press.

Lehmann, G., Harder, H.: Optical spectra of di- and trivalent iron in corundum. Am. Mineralogist **55**, pp. 98–105 (1970).

Mackowsky, M.-Th.: Die Untersuchung der Kohlen und Kokse im Auflicht. In: Handbuch der Mikroskopie in der Technik (ed. H. Freund), vol. I, pt. 2, pp. 259–328. Frankfurt: Umschau-Verlag 1960.

Mandarino, J.A.: Absorption and pleochroism: two much neglected optical properties of crystals. Am. Mineralogist **44**, pp. 65–77 (1959).

Mansberg, H.P., Kusnetz, J.: Quantitative fluorescence microscopy. Fluorescent antibody automatic scanning techniques. J. Histochem. Cytochem. **14**, pp. 260–273 (1966).

Mayall, B.H., Mendelsohn, M.L.: Errors in absorption cytophotometry: Some theoretical and practical considerations. In: Introduction to Quantitative Cytochemistry (eds. G.L. Wied, G.F. Bahr), Vol. II, pp. 171–197. New York-London: Academic Press 1970.

Mees, K.: Theory of Photographic Process. New York: MacMillan and Co. 1959.

Mendelsohn, M.L.: Absorption cytophotometry: Comparative methodology for heterogenous objects and the two-wavelength method. In: Introduction to Quantitative Cytochemistry (ed. G.L. Wied), pp. 209–214. New York-London: Academic Press 1966.

Meyer-Arendt, J.R.: Introduction to Classical and Modern Optics. Englewood Cliffs, N.J.: Prentice-Hall 1972.

Michel, K.: Die Grundzüge der Theorie des Mikroskops. 2nd ed. Stuttgart: Wissenschaftliche Verlagsgesellschaft 1964.

Monés Roberdeau, L., Bosch Figueroa, J.M., Font Altaba, M.: Quantitative study of fluorescence of brilliant cut diamonds. Fortschr. Miner. **52**, Spec. Issue: IMA-Papers 9th meeting Berlin-Regensburg 1974, pp. 521–529 (1975).

Müller, W.: Elektronische Bildauswertverfahren (Electronic Image Evaluation Methods), Microscopica Acta **71**, pp. 179–198 (1972).

Murchison, D.G.: Reflectance techniques in coal petrology and their possible application in ore mineralogy. Bull. Inst. Mining Metallurgy **73**, pp. 479–502 (1964).

Naora, H.: Schwarzschild-Villiger effect in microspectrophotometry. Science **115**, pp. 240–279 (1952).

Naora, N.: Microspectrophotometry in visible light range. In: Handbuch der Histochemie (eds. W. Gaussmann, K. Neumann), vol. I, pt. 1, pp. 192–219. Stuttgart: Fischer 1958.

Neuhoff, V.: Micro-electrophoresis on polyacrylamide gels. In: Micromethods in Molecular Biology (ed. V. Neuhoff), pp. 1–83. Berlin-Heidelberg-New York: Springer 1973.

Ornstein, L.: The distributional error in microspectrophotometry. Lab. Invest. **1**, pp. 250–262 (1952).

Ottenjann, K., Teichmüller, M., Wolf, M.: Spectral fluorescence measurements of sporinites in reflected light and their applicability for coalification studies. Lecture held at Colloque International, Pétrographie de la matière organique des sédiments, pp. 49–65. Paris: C.N.R.S. 1975.

Padgham, C.A., Saunders, J.E.: The Perception of Light and Colour. London: G. Bell and Sons 1975.

Passwatev, R.A.: Guide to Fluorescence Literature. New York: Plenum Press 1967.

Patau, K.: Absorption microphotometry of irregular-shaped objects. Chromosoma **5**, pp. 341–362 (1952).

Pelc, S.R.: Autoradiography as a cytochemical method with special reference to ^{14}C and ^{35}S. In: General Cytochemical Methods (ed. J.F. Danielli), vol. I, pp. 279–316. New York: Academic Press 1958.

Peltz, G., Heiszler, L., Hieke, E.: Optische Eigenschaften von Belegleserpapieren für die elektronische Datenverarbeitung. Das Papier. Mitteilungen aus der Zentralabteilung Technik der Siemens AG., München, 27. Jg., pp. 217–239 (1973).

Pepperhoff, W., Ettwig, H.-H.: Interferenzschichten-Mikroskopie. Darmstadt: Steinkopf 1970.

Piller, H.: Colour measurements in ore microscopy. Mineralium Deposita **1**, pp. 175–192 (1966).

Piller, H.: Influence of light reflection at the objective in the quantitative measurement of reflectivity with the microscope. Mineralog. Mag. **36**, pp. 242–259 (1967).

Piller, H.: A universal system for measuring and processing the characteristic geometrical and optical magnitudes of microscopical objects. In: Advances in Optical and Electron Microscopy (eds. R. Barer, V.E. Cosslett), vol. 5, pp. 95–114. London-New York: Academic Press 1973.

Piller, H., Gehlen, K. v.: On errors of reflectivity measurements and of calculation of refractive index n and absorption coefficient k. Am. Mineralogist **49**, pp. 867–882 (1960).

Piller, H., Prager, H.: Use of a computer in the measurement of spectral reflectances. Mineralogy and materials news bulletin for quantitative microscopic methods 1972, pp. 14–15. Cambridge (England): Polyhedron Printers Ltd. 1972.

Ploeg, M. van der, van Duijn, P., Ploem, J.S.: High-resolution scanning-densitometry of photographic negatives of human metaphase chromosomes. II. Feulgen-DNA measurements, Histochemistry **42**, pp. 31–46 (1974).

Ploem, J.S.: The use of a vertical illuminator with interchangeable dichroic mirrors for fluorescence microscopy with incident light. Z. Wiss. Mikr. **68**, pp. 129–143 (1967).

Pollister, A.W., Ornstein, L.: Cytophotometric analysis in visible light. In: Analytical Cytology (ed. R.C. Mellors), 2nd ed., Chap. 1. New York: McGraw-Hill 1959.

Preuss, J.: Mikroabsorptions- und Mikroreflexionsmessungen an organischen Molekülkristallen in Verbindung mit Röntgenstrukturuntersuchungen. Thesis, Fac. Gen. Sci. Tech. Univ., Munich 1973.

Pringsheim, P.: Fluorescence and Phosphorescence. New York: Interscience Publishers 1949. Revised edition in German. Weinheim: Chemie Verlag 1951.

Przybylski, R.J.: Principles of quantitative autoradiography. In: Introduction to Quantitative Cytochemistry-II (eds. G.L. Wied, G.F. Bahr), pp. 470–505. New York-London: Academic Press 1970.

Ramdohr, P.: Die Erzmineralien und ihre Verwachsungen. 4th ed. Berlin: Akademie Verlag 1975.

Rayleigh (4th Baron): The surface layer of polished silica and glass with further studies on optical contact. Proc. Roy. Soc., London (Ser. A) **160**, pp. 507–526, 1937.

Reeb, C.: Grundlagen der Photometrie. Karlsruhe: Verlag Braun 1962.

Rogers, A.W.: Techniques of Autoradiography. Amsterdam-London-New York: Elsevier Publ. Co. 1967.

Rosenfeld, A. (ed.): Digital Picture Analysis. Berlin-Heidelberg-New York: Springer 1975.

Rossi, B.: Optics, pp. 379–385. Reading (Mass.): Addison-Wesley Publishing Co. 1957.

Ruch, F.: The use of human leucocytes as a standard for the cytofluorometric determination of protein and DNA. In Fluorescence Techniques in Cell Biology (eds. A.A. Thaer, M. Sernetz), pp. 51–55. Berlin-Heidelberg-New York: Springer 1973.

Ruch, F., Leemann, U. (eds.): Qualitative and Quantitative Fluorescence Microscopy. Heidelberg-Berlin-New York: Springer (in preparation).

Ruch, F., Trapp, L.: A microscope fluorometer with short-time excitation and electronic shutter control. Acta Agron. Acad. Sci. Hung. **23**, 443–447 (1974).

Rudkin, G.T.: Microspectrophotometry of chromosomes. In: Introduction to Quantitative Cytochemistry (ed. G.L. Wied), pp. 387–407. New York-London: Academic Press 1966.

Sandritter, W.: Methods and results in quantitative cytochemistry. In: Introduction to Quantitative Cytochemistry (ed. G.L. Wied), pp. 159–188. New York-London: Academic Press 1966.

Schober, H.: Das Sehen (3rd ed.), vol. I, 1960, vol. II, 1964. Leipzig: VEB Fachbuchverlag.

Schrader, A.: Ätzheft. Verfahren zum Schliffherstellen und Gefügeentwicklung für die Metallographie (ed. Max-Planck-Institut für Eisenforschung, Berlin, Fed. Rep. of Germany). Berlin: Gebr. Bornträger 1957.

Sernetz, M., Thaer, A.A.: Microcapillary fluorometry and standardization for microscope fluorometry. In: Fluorescence Techniques in Cell Biology (eds. A.A. Thaer, M. Sernetz), pp. 41–49. Berlin-Heidelberg-New York: Springer 1973.

Sippl, C.J.: Computer Dictionary. Slough (Buckinghamshire): Foulsham and Co. 1966.

Smith, F.H.: A new incident illuminator for polarizing microscopes. Mineral. Mag. **33**, 725–729 (1964).

Smith, M.F.: Fiber Optics (A bibliography with abstracts). Springfield, Virginia: Nat. Tech. Inf. Serv. 1975.

Stenius, A.S.: Influence of optical geometry and absorption coefficient on diffuse reflectance values. JOSA **45**, 727–732 (1955).

Sutter, E.: Über das Spektrum von Quarz-Halogenlampen im sichtbaren Spektralgebiet. Optik **30**, 73–79 (1973).

Thaer, A.A.: Instrumentation for microfluorometry. In: Introduction to Quantitative Cytochemistry (ed. G.L. Wied), pp. 409–426. New York-London: Academic Press 1966.

Toubeau, G.: Mesure de l'absorption de la lumière dans les milieux cristallins et son intérêt en minéralogie. Bull. Soc. Belg. Géol. **LXX**, 262–282 (1961).

Trojer, F.: Reflexionsmessungen in der Mikroskopie hüttenmännischer Produkte. Berg- und Hüttenmännische Monatshefte **107**, 30–33 (1962).

Uytenbogaardt, W., Burke, E.A.J.: Tables for Microscopic Identification of Ore Minerals, 2nd revised edition. Amsterdam-London-New York: Elsevier Publ. Co. 1971.

Vašiček, A.: Optics of Thin Films. Amsterdam: North-Holland Publ. Co. 1960.

Velapoldi, J., Travis, J.C., Cassatt, W.A., Yap, W.T.: Inorganic ion-doped glass fibres as microspectrofluorimetric standards. J. Microsc. **103**, pp. 293–303 (1975).

Vos, J.C. de: A new determination of the emissivity of tungsten ribbon. Physica **20**, pp. 690–714 (1954).

Vyalsov, L.N.: Reflection spectra of ore minerals (in Russian). Moscow: The Institute of Geology, Ore Deposits, etc. Acad. Sci. USSR, 1973.

Walker, P.M.B.: Ultraviolet microspectrophotometry. In: General Cytochemical Methods (ed. J.F. Danielli), vol. I, pp. 164–217. New York: Academic Press 1958.

Walsh, J.W.T.: Photometry, 2nd ed. London: Constable and Co. 1953.

Weber, K.: Strahlungsquellen für die Mikrophotographie. Acta Histochem, Suppl. **VI**, pp. 157–163 (1965).

Weigel, O., Ufer, H.: Über Mineralfärbungen. II. Die Absorption einiger rot gefärbter Mineralien und künstlicher Präparate im sichtbaren und ultravioletten Teil des Spektrums. N. Jb. Mineral. etc. Beilageband LVII A, pp. 397–500 (1927).

Weissberger, A., Rossiter, B.W. (eds.): Physical Methods of Chemistry, Part III B, pp. 207–428. New York-London-Sydney-Toronto: Wiley-Interscience 1972.

Wendtland, W.W.: Modern Aspects of Reflectance Spectroscopy. New York: Plenum Press 1968.

Wied, G.L., Bahr, G.F. (eds.): Automated Cell Identification and Cell Sorting. New York-London: Academic Press 1974.

Witte, S.: Microscopic techniques for the in situ characterization of concentration of tissue components and penetrating molecules. Biorheology **12**, 173–180 (1975).

Yamada, S., Tsushida, R.: Polarized absorption spectra of metallic complexes and their application to the study of the structure of inorganic components. Ann. Rep. Sci. Works, Fac. Sci. Osaka Univ. **4**, 79–104 (1956).

Young, G.C.: Photomultiplier tubes and their application. EMI Document R/PC 55 T 72. Hayes (Middlesex): EMI Electronics Ltd. 1972.

Zimmer, H.G.: Geometrical Optics. Berlin-Heidelberg-New York: Springer 1970.

Zimmer, H.G.: Microphotometry. In: Micromethods in Molecular Biology (ed. V. Neuhoff), pp. 297–328. Berlin-Heidelberg-New York: Springer 1973.

Publications of Standardising Committees and Tables

Abbreviation of Names:

1. ASTM: American Society for Testing and Materials, Philadelphia, Pa. (USA).
2. British Standards Institution, London.
3. DIN: Deutsches Institut für Normung.
Papers delivered by Beuth-Vertrieb, Berlin (West).

Guide to statistical interpretation of data
   British Standard 2846: Part 1: 1975[2]

Specification for components of microscopes.
   British Standard 3836[2]

Glossary of terms used in automatic data processing
   British Standard 3527[2]

Bezugssystem der Polarisationsmikroskopie. (Reference system for polarised-light microscopy)
   DIN[3] 58879

Strahlungsphysik im optischen Bereich und Lichttechnik. (Optical radiation physics in the optical region and illuminating engineering)
DIN[3] 5031

Objektträger, Deckgläser, Immersionsmittel für Mikroskope. (Glass slides, glass cover slips and immersion oil for microscopes)
DIN[3] 58884

Lichtmessung (Photometry)
DIN[3] 5032

Optische Anschlußmaße für Mikroskope. (Optical connecting dimensions for microscopes)
DIN[3] 58887

Grundbegriffe der Meßtechnik. (Fundamental terms of measurement)
DIN[3] 1319

Bewertung und Messung der lichttechnischen Eigenschaften von Werkstoffen. (Evaluation and measurement of photometric properties of materials)
DIN[3] 5036

Recommended practice for preparation of reference with reflectance standards
ASTM[1] designation E 259-66

Standard method for absolute calibration of reflectance standards
ASTM[1] designation E 306-66

Standard method for microscopical determination of the reflectance of the organic components in a polished specimen of coal
ASTM[1] designation D 2798-72

Preparing coal samples for microscopical analysis by reflected light
ASTM[1] designation D 2797

Determination of rank by reflectance.
In: International vocabulary for Coal Petrology (ed. Intern. Comm. Coal Petrology), Supplement to: 2nd ed. Paris: C.N.R.S. 1971

Quantities, units and symbols.
The symbols committe of the Royal Society, 2nd ed. London: Royal Society 1975

International Lighting Vocabulary
Publication CIE No. 17 (E-1.1) 1970
Commission Internationale de l'Éclairage
Vocabulaire International de l'Éclairage
3ᵉ Ed. commune à la CIE et à la CEI
Bureau Central de la CIE, 4 Av. du Recteur Poincaré 75 – Paris 16ᵉ France

Klasseneinteilung und Fehlergrenzen von elektrischen Meßgeräten. (Classification and limits of errors of electrical measuring instruments) VDE 0410 18.64

American Society for Metals (A.S.M.) Metallographic Polishing. In: Metals Handbook pp. 159–162 Cleveland, Ohio: A.S.M., Reprint 1958

Subject Index

Boldface numbers refer to the most important pages for the subject.